THE

PUBLICATIONS

OF THE

SURTEES SOCIETY

VOL. CXCIII

THE

PUBLICATIONS

OF THE

SURTEES SOCIETY

ESTABLISHED IN THE YEAR

M.DCCC.XXXIV

VOL. CXCIII

FOR THE YEAR M.CM.LXXX

A SEVENTEENTH CENTURY FLORA OF CUMBRIA

William Nicolson's Catalogue of Plants 1690

EDITED BY
E. JEAN WHITTAKER

PRINTED FOR THE SOCIETY BY
NORTHUMBERLAND PRESS LIMITED
GATESHEAD
1981

CONTENTS

ACKNOWLEDGEMENTS

In the first place I would like to thank the Bishop of Carlisle, the Rt. Rev. Henry David Halsey for allowing me to visit Rose Castle and study the manuscript. I should like also to thank the Church Commissioners who permitted a microfilm and one copy to be made by the Rare Books Department, University Library, Durham to whom my thanks are also due for much help and advice. I am greatly indebted to Mr. David Crane for his critical reading of the text of the notebook and for his assistance in rendering the Latin and Greek. I have been fortunate in having been able to discuss some points with Dr. Geoffrey Halliday, Mr. C. Roy Hudleston, Dr. Edna Lind and Dr. Judith Turner. My father's knowledge of Cumbria and his generous assistance have, at all times, been invaluable.

ABBREVIATIONS

CW1	*Transactions of the Cumberland and Westmorland Antiquarian and Archaeological Society,* Old Series
CW2	*Transactions of the Cumberland and Westmorland Antiquarian and Archaeological Society,* New Series
Nicolson and Burn	Joseph Nicolson and Richard Burn, *The History and Antiquities of the Counties of Westmorland and Cumberland* (1777)
O.E.D.	*The Oxford English Dictionary*
O.S.	Ordnance Survey
PNC	*The Place-Names of Cumberland* (Cambridge, 1950–52)
PNW	*The Place-Names of Westmorland* (Cambridge, 1967)
PNY:W	*The Place-Names of the West Riding of Yorkshire* (Cambridge, 1961–63)
Synopsis 1724	ed. William T. Stearn, *John Ray Synopsis Methodica Stirpium Britannicarum Editio Tertia 1724* (Ray Society, 1973)
V.C.H.	*The Victoria History of the Counties of England*

BIBLIOGRAPHICAL NOTE

Botanical Names

The modern names for plants have been arrived at by reference to contemporary works of botany, later authorities and the appropriate local floras. John Gerard's *Herball* in the second edition enlarged and amended by Thomas Johnson, 1633 remained a most useful illustrated work throughout the century and was referred to by later herbalists up to and beyond the time of Ray. His own *Synopsis Methodica Stirpium Britannicarum* (1690), was the most complete contemporary description of the British flora and, in the later editions of 1696 and 1724, remained so well into the eighteenth century. Unless there is particular reason for referring to the earlier editions, reference has been made to the 1724 edition which is now generally accessible in the Ray Society facsimile edited by W. T. Stearn (1973). In his *Flora Anglica*, 1754, 1759, Linnaeus attempted to assign his own binomials to the Raian polynomials given in *Synopsis 1724* and a list, prepared by J. E. Dandy and printed in the Ray Society edition pages 45–62, brings these up-to-date. Many synonyms for early polynomials were also given in J. E. Smith, *The English Flora* (1824–28). In particular instances, the appropriate local flora has been consulted and a reference will be found in the footnotes. For the area of Cumbria, J. G. Baker, *A Flora of the English Lake District* (1885), William Hodgson, *Flora of Cumberland* (Carlisle, 1898), Albert Wilson, *The flora of Westmorland* (1938), F. A. Lees, *The Flora of West Yorkshire* (1888) have been used, together with Geoffrey Halliday, *Flowering Plants and Ferns of Cumbria, A Check-List* (Lancaster University, 1978).

The modern names used for flowering plants and ferns are those of A. R. Clapham, T. G. Tutin and E. F. Warburg, *Flora of the British Isles*, 2nd ed. (Cambridge, 1962).

In a few cases, although the modern equivalent of a polynomial may be beyond dispute, it may seem that the Nicolson record is unlikely. He was, however, a competent botanist and whether one should cast doubt on such records after a span of nearly 300 years is questionable. Other nomenclatural difficulties have been got over as well as possible by the use of aggregate terms and by footnoting where necessary.

Historical

Nicolson's own diaries are an invaluable source of information about his life and were published in full or in extract by the Cumberland and Westmorland Antiquarian and Archaeological Society in the *Transactions,* new series, vols i–v, xxxv, xlvi and l. There is also a very full biography of his life by F. G. James, *North Country Bishop* (Yale and Oxford, 1956), and information about his contemporaries can be obtained from Charles E. Raven, *John Ray* (Cambridge, 1942); R. T. Gunther, *Life and Letters of Edward Lhwyd* (Oxford, 1945); and R. Pulteney, *Historical and biographical Sketches of the Progress of Botany in England* (1790). Ray Desmond, *Dictionary of British and Irish Botanists and Horticulturists* (1977) is a useful source of reference for minor figures and Blanche Henrey, *British Botanical and Horticultural Literature before 1800* (Oxford, 1975) gives a most comprehensive picture of the botanical background of the age.

The following works give an account of the places where Nicolson lived, worked or visited:

C. M. L. Bouch, *Prelates and People of the Lake Counties* (Kendal, 1948).
R. S. Boumphrey, C. Roy Hudleston and J. Hughes, *An Armorial for Westmorland and Lonsdale* (Kendal, 1975).
W. G. Collingwood, *The Lake Counties*, new edition (London and NY, 1932).
J. F. Curwen, *The Castles and Fortified Towers of Cumberland, Westmorland and Lancashire North-of-the-Sands* (Kendal, 1913).
Harold R. Fletcher and William H. Brown, *The Royal Botanic Garden Edinburgh 1670–1970* (Edinburgh, 1970).
C. Roy Hudleston and R. S. Boumphrey, *Cumberland Families and Heraldry* (Kendal, 1978).
Samuel Jefferson, *The History and Antiquities of Carlisle* (Carlisle, 1838). *The History and Antiquities of Leath Ward* (Carlisle, 1840).
A. G. Loftie, *Great Salkeld, its Rectors and History* (1900).
Cornelius Nicholson, *Annals of Kendal*, 2nd ed. (London and Kendal, 1861).
J. Nicolson and R. Burn, *The History and Antiquities of the Counties of Westmorland and Cumberland* (1777).
Joseph Whiteside, *Shappe in Bygone Days* (Kendal, 1904).
James Wilson, *Rose Castle* (Carlisle, 1912).

Manuscripts

The following manuscript sources have been cited:

Ambleside, Brathay Hall Field Study Centre, Canon Hervey's botanical notebooks.
Carlisle, Tullie House, Jackson Library, microfilm and transcripts of William Nicolson's diaries. Some diaries were lost after transcription and, except for one diary in private ownership, those remaining are at Tullie House. The diaries as printed in *CW2* are incomplete and sometimes inaccurate.
Carlisle, Record Office, tithe-awards for Cumberland.
Kendal, Record Office, tithe-awards, and corn-rent-awards for Westmorland.
London, British Library, MS Harley 3780.
Oxford, Bodleian Library, MS Ashmole 1816; MS Top Gen C 27.

EDITOR'S INTRODUCTION

An Account of the Manuscript

Description

The original manuscript of William Nicolson's botanical notebook[1] is lodged in the library at Rose Castle, the episcopal seat of the Bishops of Carlisle. It was bound in calf in the nineteenth century, measures 3¾″ by 5¾″ and comprises 175 numbered pages and 179 unnumbered. It is written in Nicolson's own hand with evidence of additions and emendations over a period (see below page xlviii). Below his name on the fly-leaf, there is the date 1690, and the latest date on the manuscript is 1704 when, as Bishop, he was planting up the gardens at Rose Castle (page 115 below). Inside the front cover is the bookplate of the Carlisle See Library, Rose, and a note "Presented 1867 or 8". A note initialled H.T.W. then follows on a separate page and then, opposite, Nicolson's own mark of ownership: "Guil. Nicolson 1690." with "H. Cotton." alongside. On the same page a note, initialled by Cotton, observes:

> This volume is in the handwriting of ArchBishop Nicolson, a Cumberland man, Bishop of Carlisle, an ardent and successful student of botany, as well as history and antiquities. H.C.

Above this, there is a somewhat puzzling note in Nicolson's hand: "Nomina Herbarum ASS. vide in *E. Wanleij* Catal. MSS. p. 72, 73 &c." There seems little doubt that the work referred to is the *Antiquae Literaturae Septentrionalis Liber Alter* (Oxford, 1705), a bibliographical work in which, on pages 72 ff. is printed an Anglo-Saxon version of Apuleius Medaurensis, *De Virtutibus Herbarum.*[2] It is very likely that this is the Liber Medicinalis that Nicolson saw in Wanley's company at the Library at St. James's in January, 1702.[3] He was himself a contributor to the *Antiquae Literaturae Septentrionalis Liber Alter* and it is surprising that, when he knew and corresponded with Wanley,[4] he should have been mistaken as to the initial of his Christian name – a most uncharacteristic mistake. Moreover, when the dozen-and-a-half Anglo-Saxon names given in the *Catalogue of Plants* are compared with the list in Wanley, only three

---

1. I have, in what follows, referred to the whole manuscript as the notebook or the botanical notebook and that part which he calls *Catalogus Plantarum Britanniae*, pages 1–117, as the *Catalogue of Plants.*
2. A copy of this list will be found on pages 1–9 of John Earle, *English Plant Names from the Tenth to the Fifteenth Century* (Oxford, 1880).
3. *CW2* i 40.
4. Nicolson's letters to Wanley are on BL MS Harley 3780. The superscription is to Mr. Wanley, without initial.

(waybread, knee-holly and fennel) are common to both lists. One must, therefore, presume that this was a memorandum, made by Nicolson, and that he never went back on it to insert the Anglo-Saxon plant names in appropriate places of the *Catalogue of Plants*. Possibly he never intended to: if, as I have elsewhere suggested (page xxv below), his notebook was to be incorporated into a grander design of a history of Northumbria, it may be that the expansion of the memorandum would have been done then. It does, however, give us the last known date after which anything was added to the manuscript – 1705, the date of Wanley's publication.

Following these fly-leaf notes and the note reproduced facing p. 1, comes Nicolson's own title-page. After this come the 175 numbered pages (his pagination) of plant records, followed by an English name index which closes the *Catalogue of Plants*. Further lists and notes follow – possibly all material for the projected history of Northumbria, and these lists will be found in full or in summary on pages 100–110 below. The notebook closes with some interesting garden lists printed on pages 111–116 below.

History

An account of the manuscript's history after the author's death will be found on page xxiv of William Hodgson, *Flora of Cumberland* (Carlisle, 1898):

> It was purchased at an auction in London of Archdeacon Cotton's library, by Robert Ferguson, Esq. of Morton, at that time M.P. for the city of Carlisle. This Dr. Cotton was Archdeacon of Cashel, and Nicolson's coadjutor in the work of that diocese, which would account for the work being in the possession of the former's representatives. Mr. Ferguson, concluding that the then holder of the see of Carlisle would be the fittest custodian of the book, forwarded it in a letter dated December 29, 1868, to the late lamented Dr. Goodwin, asking him to take charge of it. The worthy prelate consented, and had the book newly and substantially bound.

One or two inaccuracies in this account require correction. The statement that Archdeacon Henry Cotton (1789–1879) was Nicolson's coadjutor in the work of the Cashel diocese is most clearly incorrect since Cotton came to Cashel almost a hundred years after Nicolson's death – and Nicolson, in any case, died in Derry. None of the work of Cashel diocese ever came under Nicolson's administration because, exactly five days after the official announcement of his appointment to Cashel, he was dead of an apoplectic stroke which killed him, according to his biographer F. G. James "as he was preparing to pack up the books and papers in his study."[5]

If that were so, it is possible that some of these books and papers had already been forwarded to Cashel at the time of Nicolson's sudden death and that the manuscript, being among them, remained there until acquired by Henry Cotton. Certainly it seems to have been Nicolson's practice to

5. F. G. James, *North Country Bishop* (Yale and Oxford, 1956), page 280.

have dealt with his books before packing up the rest of his household goods: when he moved from Great Salkeld to Rose, his books preceded his household by a week or more.[6] But if the manuscript, together with other matter, did go to Cashel in advance, then it is surprising that it stayed there and was not reclaimed by Nicolson's son John, Rector of Donaghmore in Donegal, to whom he left:

> all my Books, Scritore, Cabinet of Coins and Medals, Trunks, or Boxes of Letters, and other papers left in my Study at my death and also the Cabinet of Fossils in the office.[7]

Certainly, when John Nicolson died, only a year later, the *Catalogue of Plants* is not detailed in the "Catalogue of Books to be sold by the Administrators of the Rev John Nicholson [sic] late Rector of Donoghmore in the County of Donegal (Being the Library of the late Arch-Bishop of Cashel)", 1729, a copy of which is to be seen in the Worth Library in Dr. Steevens' Hospital, Dublin. This is, however, a catalogue of printed books and it is quite possible that the *Catalogue of Plants* was in John Nicolson's possession at his death and was dispersed through the Irish book trade to turn up, co-incidentally, in the library of Henry Cotton, Archdeacon of Cashel a hundred or so years later.

All that we can be certain of is that by 1851, the manuscript was in Cotton's possession as, in that year, there is a none-too-accurate description of it in his *Fasti Ecclesiae Hibernicae*:

> I have a small MS. volume written by [Nicolson], comprising an account of plants growing in Cumberland, and especially in the neighbourhood of Carlisle, as observed by himself in his walks.[8]

In 1868, came the sale of Cotton's library to which Hodgson refers. This was handled by Sotheby's and an auctioneer's annotated copy of their sale catalogue is now in the British Library.[9] On the third day of the sale (3rd February), the *Catalogue of Plants* is marked down as having been sold for 5/- to a buyer called Wake. A Henry T. Wake had, at about this time, become established as a bookseller in Cockermouth and he was, undoubtedly the Sotheby's buyer as there is a note, initialled by him, inside the *Catalogue of Plants*:

> From the Library of H. Cotton Archdeacon of Cashel, Author of the Typographical Gazetteer. H.T.W. 1867

The date 1867 is, presumably, an error for 1868 as there can be no doubt that the initials are Henry T. Wake's. Prior to setting up in business, he had been employed as a clerk with the firm of Suttons at Scotby, near Carlisle, and entries made by him in Suttons' ledgers (now in the Carlisle Records

6. *CW2* ii 194.
7. *CW1* iv 10.
8. Vol i (Munster), 1851, page 17.
9. *B.L.S.C. Sothebys*, 31st Jan, 1868.

Office) and, on occasion, initialled, are in the same hand as the writing and initials of the above. The last entry of his name in Suttons' books that I can find is in October, 1862 and it was, presumably, shortly after that date that he set up as a bookseller, becoming known later as an antiquarian when he published *All the monumental inscriptions in the graveyards of Brigham and Bridekirk near Cockermouth in the County of Cumberland from 1666 to 1876* (Cockermouth, 1878).

It is possible that, at the Sotheby's sale, he was acting as an agent for Robert Ferguson although the note and a marked up price of £1.11.6d inside the volume suggests that this was an independent purchase, later sold to Ferguson, who presented it to the Carlisle See Library in the manner described by Hodgson.

One last misstatement in Hodgson's description of the manuscript remains to be corrected – he states that the abbreviation S, occurring against many of the plants in the notebook, stands for *self.* In fact, as Nicolson's own explanation reproduced facing p. 1 shows, it stands for Salkeld.

Life of William Nicolson

William Nicolson was born on Whitsunday, 3rd June, 1655. Tradition, according to his biographer F. G. James, has it that he was born "with characteristic haste" in the porch of the parish church of Great Orton, where his father, Joseph Nicolson (1623–86), was the incumbent.[10] William was born the eldest in a family of three boys and four girls to a poor Cumberland parson.[11] Socially, he would be slightly better circumstanced than had been John Ray (1627–1705), the great naturalist of a generation earlier, who was the son of a village blacksmith, but financially William Nicolson, like John Ray, had his way to make in the world – his father was to die almost £400 in debt.[12] That the son became, in his forty-eighth year, Bishop of Carlisle is partly due to his Cumbrian birth – the see was traditionally held by a native of the diocese[13] – but in large part it is due to his own prodigious energy of which the *Catalogue of Plants* is but one product.

After his schooling at Dovenby, five miles from the Rectory of Plumbland, now the family home, Nicolson entered Queen's College Oxford as a batteler. It was a college with strong links with Cumberland and Westmorland, going back to the days of its founding by the Cumbrian Robert de Eglesfield, chaplain to Queen Philippa,[14] and Nicolson's own association with the college was to be a long one. Graduating in 1676, he had become known as a promising Saxonist and had been elected taberdar in the same year. In 1679, after a tour which took him to Holland and Germany under the patronage of the then Secretary of State, Sir Joseph Williamson, he became M.A. and Fellow. In the course of the tour, Nicolson had been able to visit and examine libraries and private collections and to pursue an interest in the gathering and collation of material that was to provide the impetus for much of his later scholarship. The late seventeenth century was an age of collecting and classifying and Nicolson's interest in the cataloguing of libraries and the listing of antiquities is paralleled by his interest in the collecting of fossils or the recording of plants which forms the material of the present volume.

---

10. F. G. James, *North Country Bishop, a biography of William Nicolson* (Yale, London and Oxford, 1956), page 5. In the account of Nicolson's life that follows, I have concentrated upon his interests in natural history and readers seeking a full account of his career and other interests should consult this excellent biography.
11. For a Nicolson pedigree see *CW2* i 49.
12. *James*, page 5.
13. *Ibid.*, page 3.
14. See Samuel Jefferson, *The History and Antiquities of Carlisle* (Carlisle, 1838), pages 410–14.

At twenty-four, however, Nicolson was chiefly known for his researches in Anglo-Saxon and, on his return, his patron founded a Saxon lectureship at Queen's which Nicolson was to hold for the next three years. Residual evidence of his interest in Old English is to be found in the Anglo-Saxon names which he gives to some of the plants recorded in the *Catalogue of Plants*.

At this point, it might seem that Nicolson was destined to settle down to a life of scholarship in Oxford: he was becoming something of a historian, having been asked to contribute to Moses Pitt's Atlas,[15] and there is evidence that a broad range of other intellectual interests was bringing him into contact with the most noted scholars of the age. It is likely that he had already met Robert Hooke before his departure for the continent and, on his return, he published a Latin translation of Hooke's essay "An Attempt to Prove the Motion of the Earth by Observation" (1679). Late in 1681, however, the office of Prebend of Carlisle cathedral, to be held in conjunction with the vicarage of Torpenhow, was offered to Nicolson and it seemed that he must choose between full-time scholarship and a career in the church. For a while he vacillated, being allowed to accept clerical office while, at the same time, being granted a year's grace from the college to think the matter over. In 1682, however, the college started to become insistent that he make up his mind and when, in the summer, he was offered in addition the Archdeaconry of Carlisle and the Rectory of Great Salkeld, Nicolson decided to accept.[16] A last ditch attempt to get the college to allow him to continue to hold his fellowship *in absentia* having proved abortive, Nicolson finally resigned and left Oxford in early 1683 to start a new life in the North.

In the next few years, Nicolson was to settle down in Cumberland, establishing himself at Great Salkeld and setting about his duties as Archdeacon with energy and decision. In 1686, he married Elizabeth Archer, daughter of John Archer of Oxenholme near Kendal, a substantial property owner and a notable local figure. After the wedding, the couple moved into the Rectory at Great Salkeld where they were to live in something like modest style for the next seventeen years on an income of around £200 a year.[17]

It appears to have been a life full of activity. Nicolson's work as Archdeacon took him all over the diocese and, aside from his work, there were social engagements. His diary shows him to have been welcome at Edenhall, seat of the Musgrave family for generations, and at Hutton-in-the-Forest, home of Sir George Fletcher, who was M.P. for Cumberland for

---

15. Moses Pitt, *English Atlas* (Oxford, 1680–82). Only 4 volumes of the 11 that were projected ever appeared. Volume 1 has chapters by Nicolson on Poland and Denmark, which then included Norway, and volumes 2 and 3 on the Empire of Germany were entirely compiled by Nicolson. See *James,* page 17.
16. *James*, pages 18–20.
17. *Ibid.*, pages 35–9.

virtually forty years[18] and who sent the young Nicolsons fruit trees for the Rectory garden in 1690,[19] a year in which they seem to have been particularly busy in the garden. In the summer of that year, Nicolson had undertaken several botanising trips with their neighbour from Great Strickland, Thomas Lawson and, towards the back end of the season, they appear to have persuaded him to come and help them plant up the Rectory garden with the long list of plants which will be found on pages 111–113 below. Lawson was, at that time, the most noted northern botanist of the day and it is unfortunate that Nicolson's diary gives no indication of when his friendship with him was begun. There is a gap of five years between July, 1685 and June, 1690 when he kept no diary and the first appearance of Lawson is in a characteristically terse entry on 3rd July, 1690: "Mr. Hume and I with Mr. Lawson at Trowgill and Clibburn Moss."[20] Lawson was then approaching 60, almost the same age as John Ray, to whom he had sent plant records and specimens. Born in 1630, he had been appointed in 1652 to the living of Rampside on the Furness coast just a couple of miles from modern, sprawling Barrow. 1652, however, was the year of George Fox's visit to North Lancashire and Lawson, allowing him to preach from his own pulpit, was himself converted and joined the Friends. After missionary work in the South, he married in 1659, and set up as a schoolmaster in Great Strickland.[21] Nicolson's relations with him, although cordial to the extent of exchanging visits, cannot have been too easy. It is true that the scientific world of the time was one in which there was much tolerance of religious differences. Nicolson himself opened his house to the non-conformist Leeds antiquarian Ralph Thoresby in 1694[22] and corresponded with him over a long period. But Lawson lived within the diocese and his near neighbourhood must have been something of an embarrassment to a man whose duties as Archdeacon committed him to the suppression of nonconformity within the area. None-the-less, in September, 1690, we find Lawson helping the young couple plant up their garden with a nice mixture of shrubs and border plants, wild rarities and medicinal herbs.

And there were other improvements to be made to the Rectory buildings themselves. Nicolson's predecessor, Thomas Musgrave, had done much for the property and Nicolson was to continue the improvements by extending the outbuildings and entirely reconstructing one barn.[23] From the date of

---

18. He was M.P. for Cumberland, with short breaks, from 1661 until 1700 (C. Roy Hudleston and R. S. Boumphrey, *Cumberland Families and Heraldry* (Kendal, 1978), page 116).
19. See page xxxii.
20. *Microfilm*, Tullie House Library, Carlisle.
21. Brief notices of Lawson's life will be found in *The North Lonsdale Magazine and Furness Miscellany*, Ulverston, vol i, no. 12, pages 234–7 (S. L. Petty, "Thomas Lawson, the Father of Lakeland Botany") and in *The Westmorland Notebook*, London and Kendal, vol i, pages 346–50 (M. Bennet, "Thomas Lawson, the Westmorland Botanist").
22. Ed Joseph Hunter, *The Diary of Ralph Thoresby* (1830), vol i, pages 275–6.
23. A. G. Loftie, *Great Salkeld its Rectors and History* (1900), pages 68, 104.

his marriage until the day he left "sweet Salkeld"[24] to become Bishop of Carlisle was the longest settled period of Nicolson's life and, perhaps, the happiest. In the *Catalogue of Plants*, he has left his tribute to the place with the bold "S" for "Salkeld" against plants that he found there in his walks through the lanes and fields of his home parish.

Besides a more settled way of life than he had hitherto enjoyed and a brood of eight children, Nicolson's marriage brought him relations – the agreeable Archers, whose home was always open to him as his was to them. In particular, however, it brought him a nephew – John Archer – who was to share Nicolson's interests and enthusiasms and to prove a link between remote Cumberland and the Oxford that Nicolson had so regretfully quitted.

In 1690, John Archer entered Queen's College, Oxford[25] and Nicolson was able to introduce him to the up-and-coming Welsh naturalist, Edward Lhwyd,[26] a man for whom Nicolson always had a special affection and to whom he wrote regularly over a period of many years on topics of natural history. Indeed, of all Nicolson's correspondents, Lhwyd was the one to whom he wrote in the warmest terms and the fifty-five letters[27] of Nicolson to Lhwyd preserved in Oxford, Bodleian Library, MS Ashmole 1816 reveal much more of the Archdeacon's character and temperament than do the rather impersonal entries in the diary that he kept for much of his life. Through his nephew Archer, Nicolson came to feel that he knew Lhwyd very well indeed – there is a note of friendly rivalry in "vyeing Discoveries"[28] one with another and, as late as 1701, we find Nicolson writing to Lhwyd:

> J. Archer talks of being here next month, and takeing a Ramble with me over our Mountains . . . 'Tis to good purpose to observe what plants some Counties in England produce which are never met with in Wales. 'Twas wonderful to me to find such a variety of strange faces (at no greater distance fro*m* this place then York) which I had never seen in Cumberland. You have long since told me that North-Wales and our Countrey are much of a piece. If, at Michaelmas, I hear that you have made any Discoveries, in Natural History which my kinsman and I have not noted here, I shall conclude we are Loyterers or worse.[29]

Such rambles were a regular part of Nicolson's life throughout his Salkeld days, despite a full round of clerical duties and intermittent ill-health – in June 1697 we find him writing to Thoresby:

---

24. *CW2* ii 194.
25. Joseph Foster, *Alumni Oxoniensis* (Oxford, 1891).
26. R. T. Gunther, *Early Science in Oxford, vol xiv, Life and Letters of Edward Lhwyd* (Oxford, 1945), page 125.
27. Gunther, *ibid.*, gives the number as fifty-six, but amongst them is one letter to nephew Archer.
28. Bodleian Lib, MS Ashm. 1816 fol 513.
29. Bodleian Lib, MS Ashm. 1816 fol 519.

> I have not once gotten to the top of any of our mountains this year, though I used to have rambled over a good many of them before the season was thus far advanced.[30]

How seriously we are to take his statement that he reached the top of many of the Cumbrian mountains, I do not know. He certainly climbed Blencathra (2847′) and has left us the first record of the ascent of that fell[31] and, in the essay "Botanical Apprenticeship", I have commented upon his walks on Cross Fell (2930′). Both these expeditions are made known to us through the pages of his diary. But there were, undoubtedly, other excursions that went unrecorded in the diary and which must be reconstructed from casual mentions amongst his correspondence. The fruit of these excursions is, however, surprisingly meagre. It is true that, several times when writing to Lhwyd, he remarks upon the similarity of the Cumbrian vegetation on the high tops and that of the Welsh mountains. But he does not appear to have pursued this in any systematic way and, as time went on, many of his observations in his simpling expeditions went unrecorded in the *Catalogue of Plants.* Possibly this is because, as his life grew busier, he regarded his walks as being undertaken for pure pleasure and relaxation but, more probably, it is because another scientific interest had, at the same time, taken hold upon him and was coming to engross more and more of his leisure. Lithoscopy was the science of the moment and Nicolson, like every naturalist in England, was much occupied in building up a cabinet of fossils[32] and endeavouring to answer some of the questions posed by these. Here again, it may have been the young Archer who stimulated his uncle's interest, for he was known as a keen lithoscopist and Nicolson has left us some record of what he had in his collection at Oxenholme.[33] It was, however, an interest that Nicolson was to pursue long after his interest in botany had, apparently, waned. In 1694, when Ralph Thoresby visited him, he was impressed by the "delicate collection of natural curiosities" in the Great Salkeld study[34] and, as late as 1709, in Nicolson's last letter to Edward Lhwyd, a couple of months before the latter's sudden death in June, we find him asking for fossil duplicates so that he might put his own collection into an order modelled upon Lhwyd's system.[35]

As with his botanising, however, it seems that Nicolson's interest in fossils was largely that of a collector and recorder, rather than that of a

30. Ed Joseph Hunter, *Letters of Eminent Men addressed to Ralph Thoresby* (1832), vol i, page 292.
31. E. Jean Whittaker, "The First recorded ascent of Blencathra", *Cumbria*, Feb, 1978, pages 622–4.
32. This was eventually left to his son, John. Nicolson's will is reproduced *CW1* iv 9–10.
33. Archer contributed to Edward Lhwyd's *Lythophylacii Britannici ichnographia* (1699), and some lists of his specimens are to be found in the Tullie House (Carlisle) transcripts of Nicolson's diaries, vol i, pages 135–6.
34. Ed Hunter, *Diary of Thoresby*, vol i, page 275.
35. *James*, page 204.

speculative philosopher. Like any other naturalist at the end of the seventeenth century, he took an interest in the problems that fossils posed to the Mosaical account of Creation, but his opinions were aired only in his correspondence and he never ventured into print to add to the increasingly acrimonious debate that revolved, in the closing years of the century, around the vexed question of Noah's Flood. In 1696, however, he was drawn unwillingly into the arena when his less reticent neighbour, Thomas Robinson, Rector of Ousby, published a work entitled *New Observations on the Natural History of this World of Matter and this World of Life.*

The prefatory letter to this work is addressed to Nicolson and conveys the impression that he had seen and endorsed all the opinions expressed by Robinson in the succeeding pages. Its tone is an unfortunate mixture of self-assurance and patronage of those who did not hold Robinson's eccentric opinions and, in the text that follows, he goes on to tilt recklessly at all his eminent predecessors in the field and, in particular, at John Woodward, author of the *Essay toward a Natural History of the Earth* (1695). Woodward was an ill man to cross at the best of times and the resultant row can have done little to encourage Nicolson to venture into print with any reflections of his own on the subject of the Creation and the Flood – the more especially as he was obliged to spend a good deal of ink and paper denying that he had had any foreknowledge of the contents of Robinson's book.

During these busy years, Nicolson was, in addition, contemplating a work which would have drawn together his eclectic interests in botany, geology, language, history and antiquities. This was his "*Northanhymbraric* or Description of the antient Kingdome of Northumberland" the uncompleted manuscript of which is to be found on MS Top Gen C 27 in the Bodleian Library. What form this work would have finally taken, we cannot, of course know. The existing manuscript is in Latin, but the times were not propitious for publications in Latin[36] and it may be that the final result would have been closer to a work like Plot's *Oxfordshire* (1677), or his *Staffordshire* (1686), which were popular at the time and very well thought of. Certainly topographical/historical works were in demand as the times moved on towards the eighteenth century – Nicolson himself offered to subscribe to a work of John Aubrey's,[37] possibly the *Natural History of Wiltshire* which, like Nicolson's own Natural History was destined to remain unpublished by its author.[38] The uncompleted drafts of both men

36. Booksellers were becoming increasingly reluctant to undertake learned works and some account of the difficulties experienced by Lhwyd in getting his important work on British fossils, the *Lithophylacii Britannici ichnographia* through the press in the late 1690s is to be found in Melvin E. Jahn, "A Note on the editions of Edward Lhwyd's *Lithophylacii Britannici ichnographia*" (*Journal of the Society for the Bibliography of Natural History*, vol vi, 1971–74), pages 86–97.
37. Letter of Nicolson to John Archer, 5th June, 1693, *Bodleian Lib, MS Ashm.* 1816 fol 466.
38. This was eventually published, edited by J. Britton, by the Wiltshire Topographical Society in 1847.

were, however, used in part in the compilation of Edmund Gibson's edition of Camden's *Britannia* in 1695. I have dealt separately with Nicolson's botanical contribution to this work,[39] but it should be noted here that in the Northumberland and Durham sections of the *Britannia* Nicolson made considerable antiquarian contributions – in which field he was, indeed, more noted. His own work, however, remains unpublished, but it seems likely that his notebook would have been used to complete it. There is virtually no material relating to natural history in MS Top Gen C 27, despite his stated intention of having a natural history section in the work.[40] Notes on fishes, birds and curiosities of the north at the end of the notebook, however, suggest that this may have been regarded as the reservoir from which he was to draw up his account of the flora and fauna of the Northern counties. All these, however, were but the preliminary notes for a book that never was. Nicolson's nephew, Joseph, in collaboration with Richard Burn, was to produce the first county book of Westmorland and Cumberland – the familiar *Nicolson and Burn* published in 1777. But William Nicolson found that life and his career caught him up and carried him forward to greater honours than he had had any cause to expect. In 1703, he was offered the Bishopric of Carlisle and, when he accepted, any chance than he might have had of completing such an ambitious work of scholarship as he had planned, was postponed as the obligations of office closed around him.

There can be no doubt that Nicolson enjoyed his new position. Always an able and decisive administrator, he was now in a place of influence and power. His life became busier than ever and, although he kept in touch with the world of scholarship, there seems to have been little time for sustained work on his own account. The business of the diocese, his immediate and extended family and the ramifications of local politics: all these conspired to fill his life. But he continued to write to Lhwyd and he encouraged younger scholars, attempting for instance to found permanent lectureships in Anglo-Saxon at both Oxford and Cambridge.[41] Moreover, as Bishop, he now made regular journeys to London to take his place in the House of Lords and his diary is scattered with the names of the scholars and virtuosi that he met on these occasions. In November, 1705, the unpredictable Woodward proposed him for a Fellow of the Royal Society and he was duly elected.[42] The following week, having been admitted by Sir Isaac Newton, then president, we find him attending a meeting and demonstrations at Gresham College where the Society was gathered to hear, amongst other things, Hans Sloane read a letter dealing with an interesting medical case at Harwich. Such visits kept alive Nicolson's interest in scholarship and

---

39. See pages xxxix–xlvi below.
40. In a letter of 23rd September, 1691, he sent a scheme of the work to Ralph Thoresby (Hunter, *Correspondence,* vol i, page 116) and this scheme is reproduced in a draft letter to Gibson forming a contents page to Bodleian Lib. MS Top Gen C 27/2: "A short acct. of ye Title, Design, &c. of my *Northumb*. sent to Mr. Gibson Jul. 23. 94."
41. *James*, page 226.
42. *CW2* iii 33.

discovery and, back at home, he was still collecting fossils and geological specimens and arranging and recording these. It seems unlikely, however, that he added to the *Catalogue of Plants* after the century had turned and he became Lord of Rose.

In 1718, there came a new move. The Bishopric of Derry fell vacant and Nicolson, mindful of his family's needs, accepted it when it was offered to him. Carlisle had always been amongst the poorer bishoprics and Ireland offered opportunity for settling relatives and family. In 1719, Nicolson left Cumberland to take up his new duties, returning only once (in 1722) for a last visit.[43]

In 1727, a new honour – the Archbishopric of Cashel was to be Nicolson's but, only five days after his translation became official, Nicolson was dead of an apoplectic stroke.[44].

So, on 14th February, 1727, at the age of 71, died William Nicolson, generally referred to since as "Bishop Nicolson" – a style of address that lays the emphasis squarely where it belongs: on the man in his public and episcopal function. Nicolson made his own choice in 1682, when he gave up his Oxford fellowship – thereafter, all his scholarship was but the ornamenting of an active life as Archdeacon and ruling Bishop.

That having been said, as it must be, we are still left with an impressive record of achievement. Nicolson is known to most people interested in Cumbria as a historian and antiquarian – the work in front of us will demonstrate his capacity as a botanist. On pages xlvii–li below, I have tried to arrive at some assessment of that capacity, but that is only a small part of his talent. He was thorough, enquiring, late-seventeenth-century man – a Royal Society man, but one who was amongst the later heirs of the Renascence tradition of man of affairs and man of intellectual accomplishments. But above all, perhaps, he was a Cumbrian. Writing to Humphrey Wanley from Rose in 1705, he said:

> Next to what concerns the preservation of our establish'd Religion and Government, peace here and salvation hereafter, I know nothing that has a greater share in my Thoughts and Desires than the promotion of the Septentrional Learning.[45]

---

43. *James*, pages 266–8.
44. *Ibid*., page 280.
45. BL MS Harley 3780 fol 269.

1690, the year that Nicolson opened his botanical notebook, was the year that John Ray gave British botanists their first pocket flora – the *Synopsis methodica Stirpium Britannicarum*. With a second edition in 1696 and a third in 1724, this was to be "the pocket companion of every English botanist"[46] for nearly a century after its first publication. There is hardly a copy to be found today without the marginal additions and comments of some observant user and its own pages reveal, better than anything else can, the network of botanists and "moss croppers"[47] who sought to cover the British Isles and sent in their records to the cottage at Black Notley where John Ray worked to give British botany "a secure and intelligible foundation."[48] Their names are scattered in the text and the exactness with which they record plant locations testifies to an enthusiasm that Ray himself was largely responsible for creating. And beyond these again, are a small army of helpers, whose names, if they appear at all, receive only the barest of mention. We know from Nicolson that Thomas Lawson sent a man over to Walney Island from time to time to gather maritime plants[49] and a friend of Lawson's, the Quaker Reginald Harrison, finds his way into Camden's Britannia as the first discoverer of *Serratula tinctoria.*[50] Of such enthusiasts, we know only the names and these often second-hand through such botanists as Edward Lhwyd who gathered together a band of occasional helpers who were paid by results and rewarded for interesting finds. Some, indeed, became pensioners and were provided for out of a fund to which Nicolson was a subscriber.[51] One or two of these collectors were to undergo hardships and adventures that foreshadow those of later expeditionaries consumed by the specimen-hunting fever: in April, 1693, we find Lhwyd worrying about the fate of a shoemaker-Lithoscopist whom he fears lost "in the late great snow,"[52] and his own safety was from time to time to cause concern to his friends, including Nicolson who we find anxious that his friend may be lost in "one or other of the Cambrian Mountains."[53] Such fears were not

46. William T. Stearn, quoting this description from Pulteney's *Historical and biographical Sketches of the Progress of Botany in England* (1790), compares the *Synopsis* with the Floras of Bentham and Hooker and Clapham, Tutin and Warburg and points out that it fulfilled the same function of field guide and source of reference (*Synopsis 1724*, page 3).
47. See below n 93 page 50.
48. Charles E. Raven, *John Ray* (Cambridge, 1942), page 259.
49. Ed John Nichols, *Letters on Various Subjects, Literary, Political and Ecclesiastical to and from William Nicolson 1683–1726/7* (1809), vol i, page 110.
50. See below page xxxix.
51. Gunther, *Life and Letters of Edward Lhwyd*, page 8.
52. *Ibid*., page 177.
53. Bodleian Lib, MS Ashm. 1816 fol 498.

entirely without foundation as Lhwyd, in his native Wales, had been alternately suspected of being a Jacobite spy or a tax-gatherer and when, in 1700, he visited Ireland, he reported ruefully that "the Tories of Kil-Arni in Kerry obliged us to quit those mountains much sooner than we intended."[54]

In remote Cumberland, no such adventures seem to have befallen Nicolson in what he describes, undramatically enough, as his "rambles" with his nephew John Archer. Together they roamed the Cumbrian fells at such times as nephew Archer's university vacations co-incided with Archdeacon Nicolson's opportunities for leisure armed, very likely, with some such simpling book as Lhwyd recommends to David Lloyd at Blaern-y-ddol:

> You must have a pretty large simpling Book with a stif cover; & be sure of half a dozen patterns of each plant you meet with on these high Hills, in what posture soever you finde 'm.[55]

Some such simpling book Nicolson probably had – before he started *Catalogue of Plants*, he was using his diary for botanical records (see below pages xxx–xxxviii) and the entries therein are always in alphabetical order. It is reasonable to suppose, therefore, that the day's specimens lay before him when he was writing up and he noted them down in order as he examined them. The manuscript of the *Catalogue of Plants*, indeed, has a few leaves pressed between its pages and it is pleasant to suppose that these are relics of the Archdeacon's simpling expeditions round and about Cumbria.

Not all his records, of course, were his own. His Cumbrian associates Thomas Lawson and John Robinson receive acknowledgement and, besides records, he would undoubtedly have received specimens of plants from fellow enthusiasts. John Robinson, we know, sent him a specimen of the delicate fly orchid, *Ophyrs insectifera*, for Nicolson notes in his diary his pleasure at receiving this. (See below page xxxii.)

Contemporary herbaria such as those of Sloane and Morison reveal something of this traffic in specimens and we may turn again to Lhwyd to see how plants were packed for despatch in the days of the Carrier's cart:

> You must get a box of an indifferent size; such as you might guesse would scarce contain them; then lay in some mosse at the bottom of it lightly besprinkled with water; soe lay in the shrubs & greater plants first pressing them down with your hands pretty close; then a little mosse lightly wetted; and soe the rest of the plants, putting here & there alike mosse upon them as you lay 'm in. When all are put in fill up the box with mosse, that they may have noe room to be disturbed on the carriage & besprinkle it lightly with water soe nayl it up securely, borring some small holes in several parts of the cover, wherein the carrier must besprinkle a handfull or two of water every night; and see the box layd in a Seller or some cool place.[56]

---

54. *Gunther*, page 33.
55. *Gunther*, page 69.
56. *Gunther*, page 73.

One can only hope that if Nicolson forwarded any of the "Sets, Seeds, Fruits and Specimens" asked of him by the Scottish botanist, James Sutherland, in 1701, that he found so conscientious a Carrier. There is a rueful note in Nicolson's listing of the garden plants sent by Sutherland, perhaps by way of reciprocation, in 1703:

> "... Shrubs, sent from Mr Sutherland; Or, at least, so many of 'em as the Carrier brought to my hand."[57]

Whether or not he kept a herbarium, Nicolson does not say – he certainly shared the general collecting proclivities of the age – but in 1690, after a summer of expeditions, he sat down and began a record that would be more resistant to insect pests and fungal decay. From that date on, the *Catalogue of Plants* became the repository for information gathered on the Archdeacon's botanising expeditions. It is even possible that it accompanied him around his home parish as a field notebook. When he writes "S" for Salkeld against a plant, he almost always uses plummet – only a couple have been written in ink, the rest present a waxy, reddish appearance and are written very bold. It seems at least possible that the abbreviation was jotted down as and when Nicolson noticed a new plant in walking or riding around his parish. He would hardly carry an ink-horn on such occasions and the small notebook and a piece of plummet would fit very handily into a pocket.

If he should wish, back at home, to consult some work of identification, there were now many available to him. If we look at the list of authorities and references which occur in the text, we can trace, in brief the history of botany as Nicolson knew it. The ancients are still there, represented by Theophrastus and Dioscorides whom Thomas Johnson, when he prepared the second edition of Gerard's Herball, set to adorn the portals of his title-page engraving. Then there are the continental botanists of a hundred years earlier, de l'Ecluse and Matthioli, Dodoens and de l'Obel. From these men, the slower-starting British botanists drew their information and they, in turn, are well represented in Nicolson's *Catalogue* – the great herbalists Gerard and Johnson and Parkinson, followed by a whole generation of post-restoration students of plant-life who were moving away from the herbal tradition to found a more truly scientific botany. To this group Nicolson, with his precise and careful records, may fairly claim to belong. The next essay goes on to see how he set about training himself in the observation and recognition of plants.

---

57. See below page 114.

Botanical Apprenticeship

There is little early evidence of William Nicolson's interest in botany – his diaries for 1684 and 1685 contain no mention of plants and for five years thereafter he kept no diary. In the summer of 1690, however, he became, if he had not been so before, a keen plant hunter. The first recorded expedition of his takes place in early June and others followed during that warm summer.

First excursion: Eden Valley, 6th June, 1690

This first recorded expedition took place in the meadows and pastureland about Salkeld, Langwathby and Edenhall. Nicolson was accompanied by "Hume" and "Dawes". Dawes is mentioned again by Nicolson and would appear to have been his brother-in-law's nephew, Thomas Dawes of Barton.[59] The other partner in this first excursion – "Hume" – would be Robert Hume, later Vicar of Aspatria, but at this time the incumbent at Lazonby, who appears intermittently in the diary going about the work of the diocese with Nicolson.[60] With these two, Nicolson observed the following plants and entered them in his diary afterwards:

1. Agrimonia vulgaris. (*Agrimonia eupatoria* L.)
2. Bistorta minor. (*Polygonum bistorta* L.)[61]
3. Daucus. (*Daucus carota* L.)
4. Fumaria. (presumably *Fumaria officinalis* L.)
5. Lychnis Sylvestris, flore albo. (*Silene alba* (Mill.) E. H. L. Krause)
6. Orobus Sylv*aticus* nostras. (*Vicia orobus* DC.)
7. Trachelium minus. (*Campanula glomerata* L.)
8. Trifolium Luteum odoratum. (*Melilotus altissima* Thuill.)
9. Vicia multiflora maxima. (*Vicia sylvatica* L.)

Vicia orobus, we may notice, had been one of the more interesting finds of the great John Ray when he had come north in the 1660s[62] and found it not far from here at "*Gamlesby* in *Cumberland*, about six Miles from *Pereth*."[63]

---

59. *CW2* xxxv 102: "Br *Nevinson*, bespeaking Entertainmt for his Nephew *Dawes*." Thomas Dawes married Elizabeth, daughter of John Nevinson, who was Thomas Nevinson's eldest brother. And Thomas Nevinson was Brother Nevinson to Nicolson by virtue of the fact that he married Nicolson's youngest sister, Grace. For pedigree see *Nicolson and Burn*, vol i, page 452 (Nevinson) and *CW2* i 49.
60. See *CW2* i 26, 27, 30.
61. See below n 32 page 17.
62. See Raven, *Ray*, pages 113–20.
63. *Synopsis 1724*, page 324.

Second excursion: Visit to Hutton, 12th June, 1690

Alongside his awakened interest in botany, Nicolson, who was now comfortably settled at Great Salkeld, appears at about this time to have started to take an interest in gardens and gardening. Gardening was, in any case, much in discussion as the seventeenth century moved towards the Augustan times of the early 1700s and towards the Horatian ideal of man the estate manager. The last decade of the seventeenth century saw the gardens at Levens begun under the direction of Monsieur Beaumont[64] and, when Nicolson visited Sir George Fletcher's garden at Hutton-in-the-Forest, he took note of garden plants as well as interesting natives. His list headed British Plants comes first:

1. Althaea vulgaris. (*Althaea officinalis* L.)
2. Angelica Syl*vestris* minor. (*Aegopodium podagraria* L.)
3. Antirrhinum majus. Alb*um* et Purp*ureum*. (*Antirrhinum majus* L.)
4. Atriplex olida. (*Chenopodium vulvaria* L.)
5. Carduus mariae vulgaris. (*Silybum marianum* (L.) Gaertn.)
6. Euonymus Theophrasti. (*Euonymus europaeus* L.)
7. Helleborus Niger Hortensis, flore viridi. (*Helleborus viridis* L.)
8. Lepidium Latifolium. (*Lepidium latifolium* L.)
9. Lilium Convallium, Album. (*Convallaria majalis* L.)
10. Lingua Cervina officinarum. (*Phyllitis scolopendrium* (L.) Newm.)
11. Persicaria Siliquosa. (*Impatiens noli-tangere* L.)
12. Rhodia Radix. (*Sedum rosea* (L.) Scop.)
13. Rhus myrtifolia Belgica; seu, Elaeagnus Cordi. (*Myrica gale* L.)
13\. Scordium. (*Teucrium scordium* L.)
14\. Solanum Lignosum. (*Solanum dulcamara* L.)
15\. Pentaphylloides fructicosum. (*Potentilla fruticosa* L.)

These were the wild rarities of the garden, and Nicolson followed this list by another of interesting cultivated plants ("Rariores Hortenses"). Salad herbs such as Orache (number 5) and Skirrets (17) and the apothecaries' Asarabacca (3) and Masterwort (11) are noticed alongside showy bulbs such as the "Asphodelus Luteus" (4) and the Martagon type Lily (12). Gold of Pleasure (13) apparently counts for a rarity as does the undoubtedly dramatic but native Monkshood (1) and the familiar Lords-and-Ladies (9). The plants of the New World are represented with a type of Everlasting (10), Tobacco (15), American Daisies (8) and Nasturtiums (14), which were familiar to Parkinson at the beginning of the century when he wrote in cautionary vein:

> This faire plant spreadeth it selfe into very many long trayling branches ... whereby it taketh up a great deale of ground."[65]

64. W. G. Collingwood, *The Lake Counties*, new edition (London and NY, 1932), page 13.
65. John Parkinson, *Paradisi in Sole Paradisus Terrestris 1629* (facs. ed. 1904), page 280.

Lastly, we might notice the trees and shrubs – a Horse Chestnut (6) followed by a type of Senna (7), a Berberis at number 16 and the white and blue lilacs at number 18. Here is the complete list:

1. Aconitum Coeruleum.
2. Apium dulce. *Selery.*
3. Asarum.
4. Asphodelus Luteus.
5. Atriplex Hortensis, pallide virens.
6. Castanea Equina.
7. Colutea Scorpioides.
8. Doronicum Americanum.
9. Dracontium majus.
10. Gnaphalium Americanum.
11. Imperatoria.
12. Lilium montanum minus.
13. Myagrum Sativum.
14. Nasturtium Indicum majus.
15. Nicotiana minor.
16. Oxyacantha Dioscoridis seu Theophrasti.
17. Sisarum.
18. Syringa. Coerulea, et Alba.
19. Urtica *non* urens, folio variegato, ex Insula Canada.

The orchard does not receive a mention here but at this or some other time Nicolson must have been promised some fruit trees from Hutton for, on November 19th, we find him recording that J. Smith, gardener, brought him from Hutton Apricock, Kentish Cherry, Black Heart, Plums, Vines.

Third excursion: John Robinson's garden, The Ghyll, near Kendal, 18th June, 1690

At this time, Nicolson must have already been known to be an interested student of botany as, on 10th June, he had been sent a specimen of "Orchis Myodes" (*Ophrys insectifera* L.) from Dr. Robinson of Kendal – "planta nostras perelegans" says the delighted Nicolson. John Robinson, who is sometimes called Fitz-Roberts,[66] had already achieved recognition as a botanist; his records were to be quoted in the second edition of John Ray's *Synopsis*, which came out in 1696, and he was to be well thought of by Petiver and Doody and those who sought to build on Ray's foundation. He appears to have lived at the house called The Ghyll, a former farm-house just above Kendal on the Brigsteer road and, on 18th June, when it happened that he was in Kendal to preach, Nicolson visited his garden and made a list of things of interest:

1. Acetosa Rotundifolia West*morlandica*. (*Oxyria digyna* (L.) Hill)
2. Acorus verus. (*Acorus calamus* L.)
3. Betonica Aquatica. (*Scrophularia aquatica* L.)
4. Conysa major. (*Inula conyza* DC.)
5. Conysa media. (*Pulicaria dysenterica* (L.) Bernh.)
6. Cotyledon Hirsuta. (*Saxifraga stellaris* L.)
7. Geranium Nodosum. (*Geranium nodosum* L.)

---

66. See Ray Desmond, *Dictionary of British and Irish Botanists and Horticulturists* (1977), page 525.

8. Geranium fuscum.
9. Geranium longius radicatum.[67]
10. Marrubium Aquaticum. (*Lycopus europaeus* L.)
11. Pentaphyllum erectum folio Argenteo. (*Potentilla argentea* L.)
12. Polygonatum.
13. Ptarmica Vulgaris. (*Achillea ptarmica* L.)
14. Serratula folijs *non* dissectis. (*Serratula tinctoria* L.)
15. Tapsus Barbatus. (*Verbascum thapsus* L.)

Perhaps there was peace to be found in the contemplation of such a garden, for the times were not easy – the country was only just adjusting to the rule of Dutch William and non-juring ministers were being ejected wholesale. On this occasion, Nicolson took as the text of his sermon Joel 2:17 "Give not thine heritage to reproach that the heathen should rule over them." In Nicolson's mind, the heathen was but one remove from the Scot and the Jacobite and he himself had taken the new oaths rather than let in such to rule over the country. Perhaps if he had been less of a Cumbrian, with a long memory of trouble from Scotland, or less of a Church of England man, alarmed by the growing Popery of the Stuart succession, he might have renounced his career on a scruple as John Ray had done in 1662 when he had found the Act of Uniformity quite unacceptable.[68] By that decision, Britain gained one of her greatest naturalists. Nicolson, forced in 1689 to choose, chose, as he had done at Oxford, the church and his career.

It was a choice that he was prepared to defend. In 1715, when the Jacobite armies crossed the border, Nicolson was amongst the force that sought to intercept them on Penrith Fell. When this force melted away at the sight of the Highland army, the story goes that the Bishop, as he now was, was prepared to take the entire army on single-handed had not his coachman, being of less heroic stature, whipped up the horses for a rapid return to Rose.[69] Thirty years later, the retreating forces of another Jacobite army made history just a little south of Penrith where, on Clifton Moor, the last engagement on English soil took place – a ground familiar to Nicolson where he had noticed the Knotted Pearl-wort, *Sagina nodosa*.

Fourth excursion: down River Kent and back through Sedgwick and Natland 19th June, 1690

The following day, he was still in Kendal, taking advantage of his stay to enjoy a short trip down the River Kent and back through Sedgwick and Natland. John Robinson accompanied him – perhaps as the local expert – and the two were joined in their exploration of the unfrequented byways by

---

67. See below n 94 page 103.
68. Raven, *Ray*, pages 60–61.
69. C. M. L. Bouch, *Prelates and People of the Lake Counties* (Kendal, 1948), pages 304–7.

a Dr. Sutch M.B., who is described as a Kendal doctor, and "A. Farrington", who would be Alexander Farrington, then Master of Kendal Grammar School, who later became Vicar of Penrith and died there in 1699. In his diary Nicolson mentions visits to a Mrs. Farrington in Kendal, who may have been Alexander's widow.[70] Indeed, Nicolson may have been connected with the family through his wife – he uses the term "affinis"[71] to describe him, a term generally used of a relation by marriage. It was to be a day notable for marine fossils,[72] found in the mill dam above the force at Levens, for *Meum athamanticum* "in ye Fields near Kendal-Castle" and for "Euonymus Theophrasti" (*Euonymus europaeus* L.) "exceedingly plentiful in ye Hedges about Natland and Sedgwick." Meum was, indeed, a local speciality, known in South Westmorland from the time of Gerard, who observes that the plant was to be found at "Round-thwait betwixt Aplebie and Kendall."[73]

Fifth excursion: Cross Fell, 2nd July, 1690

Conditions in Cumbria during that hot summer of 1690 must have been particularly happy for botanising. Down in London, the diarist John Evelyn noted that it had been a "very extraordinary fine season"[74] and, on 2nd July, the energetic Nicolson is to be found climbing on Cross Fell "with S*ir* C. M. and Mr. Hutton." Sir Christopher Musgrave, then 58, had become the M.P. for Westmorland in the recent election of 1689–90 while his son, also Christopher, had replaced him as the M.P. for Carlisle.[75] During his career Nicolson's political affinities were to swing away from the Tory Musgraves and towards the Whig Lowthers,[76] but this was to be a day enjoyed upon the fells in the company of Sir Christopher and one of Sir Christopher's Hutton relatives. The Musgrave party were, however, keener on bilberry-picking than botanising[77] and, whilst they did so, Nicolson collected plants, noting the pretty Cloud Berry for which Cross Fell is noted and, in the evening, entering up a clubmoss and two lichens in his diary:

1. Muscus erectus Abietiformis. (*Lycopodium selago* L.)
2. Muscus Coralloides apicibus coccineis.
3. Muscus Pyxioides.

---

70. *CW2* ii 211, *CW2* iii 56, for instance.
71. Wrongly printed as "affiuis" in *CW2* i 34.
72. "Lapides Entrochi in unam massam concreti." *CW2* i 34.
73. John Gerarde, *The Herball or Generall Historie of Plantes* (1597), page 895.
74. Entry for 3rd August.
75. R. S. Ferguson, *Cumberland and Westmorland M.P.'s* (London and Carlisle, 1871), page 65.
76. *James*, pages 41 ff. and 170 ff.
77. "Gallis palustribus (Aldrovandi, *morennis* Anglorum) insidiebant; ego plantis; et speciatim *Chamaemoro* quo ibi frequens," writes Nicolson in his diary, and although the former are strictly Bog Whortleberries (*Vaccinium uliginosum*), Nicolson probably just meant bilberries in general.

Sixth excursion: Trowgill and Cliburn Moss, 3rd July, 1690

The fine weather was to last all July, and the day following the Cross Fell expedition saw Nicolson again out-of-doors – this time in the neighbourhood of Cliburn at "Trowgill" (Trough Gill: NY 587240). Nicolson's companions this time were Mr. Hume of the 6th June expedition and Mr. Lawson – the first mention of Thomas Lawson in Nicolson's diary. The party chose their ground to concentrate particularly upon the plants of the bogland:

1. Asphodelus Lancastriae. (*Narthecium ossifragum* (L.) Huds.)
2. Eupatorium folio Integro. (*Bidens cernua* L.)
3. Filix Aquatica. (?*Dryopteris cristata* (L.) A. Gray)
4. Gramen Junceum Leucanthemum. (*Rhynchospora alba* (L.) Vahl)
5. Gramen Leptopursacaulon. T.L.
6. Gramen Triglochin. (*Triglochin palustris* L.)
7. Hypericum pulchrum Tragi. (*Hypericum pulchrum* L.)
8. Millegrana minima. (*Radiola linoides* Roth)
9. Muscus Polyspermus.
10. Nummularia flore purpureo. (*Anagallis tenella* (L.) L.)
11. Sparganium Ramosum. (*Sparganium erectum* L.)
12. Veronica Mas. (*Veronica officinalis* L.)

Seventh excursion: Thomas Lawson's garden, 7th July, 1690

The discovery of common botanical interests was to bring Lawson and Nicolson together again. On 7th July, we find Nicolson visiting Lawson's garden at Great Strickland with Dr. Sutch and Mr. Robinson. Dr. Sutch is the Kendal doctor whom we have met already and Mr. Robinson is not Dr. John of Kendal, but Thomas Robinson, Nicolson's neighbour at Ousby. Indoors the visitors were shown a sea-weed, "Alga latifolia porosa, maxima", but the main interest was the garden, with its mixture of wild and cultivated rarities:

1. Aristolochia rotunda. (*Aristolochia clematitis* L.)
2. Cerasus Sylv*estris* Septentrionalis fructu parvo rubro Serotino. (*Prunus cerasus* L.)
3. Chamaemespilus.
4. Epimedium.
5. Fagus. (*Fagus sylvatica* L.)
6. Geranium Haematodes, flore eleganter variegato. (*Geranium sanguineum* L. var lancastrense (With.) Druce)
7. Hypericum frutescens. (*Hypericum androsaemum* L.)
8. Jesminum Persicum.
9. Periclymenum perfoliatum.
10. Populus Alba. (*Populus alba* L.)
11. Ruscus. (*Ruscus aculeatus* L.)
12. Pseudo Cytisus. (*one of the European broom-like plants*)
13. Staphylodendron. (*Staphylea pinnata* L.)

Eighth excursion: Mayburgh near Penrith, 10th July, 1690

Three days after the visit to his garden, Lawson joins Nicolson in a visit to the archaeological site at Mayburgh or, as Nicolson spells it, Maburg.[78] This, presumably, was an outing initiated by Nicolson who, with his keen interest in archaeological matters, was well qualified to speculate on the origins of the remains and to show the older man round the site. In return, it was probably Lawson who led the way in the search for plants about the banks of the River Eamont. Afterwards, as usual, Nicolson entered these up in his diary:

1. Allium montanum purpureum proliferum. At *Maburg*. (*Allium carinatum* L.)
2. Androsaemum Hypericoides. (*Hypericum hirsutum* L.)
3. Gentianella flore lacteo.
4. Geranium Saxatile lucidum. (*Geranium lucidum* L.)
5. Lilium Convallium. (*Convallaria majalis* L.)
6. Lysimachia purp*urea* Spicata. (*Lythrum salicaria* L.)
7. Marrubium Aquaticum. (*Lycopus europaeus* L.)
8. Saxifraga Anglica palustris. (*Sagina nodosa* (L.) Fenzl)

Ninth excursion: Gardens at Lowther and Strickland, 28th July, 1690

The last outing of the month saw Nicolson again gathering instruction from gardens. He and Dr. Law, who was to become Nicolson's family physician,[79] joined Lawson on a visit to gardens at Lowther and Strickland. The following list is the result:

1. Acer minus. (*Acer campestre* L.)
2. Alnus nigra baccifera. (*Frangula alnus* Mill.)
3. Aria Theophrasti. (*Sorbus aria* agg.)
4. Carduus moschatus. (*Carduus nutans* L.)
5. Cerasus Sylv*estris* fr*uctu* parvo cordiformi. (*Prunus cerasus* L.)
6. Cornus foemina. (*Thelycrania sanguinea* (L.) Fourr.)
7. Glycyrrhisa Sylv*estris*.
8. Helleborine flore albo. (*Cephalanthera damasonium* (Mill.) Druce)
9. Lathyrus Sylv*estris* maximus. (*Lathyrus sylvestris* L.)
10. Lysimachia Lutea. (*Lysimachia vulgaris* L.)
11. Mentastrum.
12. Myagrum. (*Camelina sativa* (L.) Crantz)
13. Pneumonanthe. (*Gentiana pneumonanthe* L.)

---

78. See below n 8 page 9.
79. Robert Law, Doctor of Physick, was, in 1708 if not earlier, living at Chalk Hall or Chalkfoot, Dalston, called by Nicolson "Shauk" (*PNC* i 151), not Shank as printed in *CW2* xxxv 120. Law later removed to Newcastle, possibly between May and August 1712, when he was replaced as the family physician by Dr. Archer, who attended Nicolson's wife in her last illness (See *CW2* iv 58, 59). Law's will was proved at Durham on 14th March, 1727.

14. Verbascum nigrum. (*Verbascum nigrum* L.)
15. Viorna. (*Clematis vitalba* L.)

Tenth excursion: Cross Fell, 7th August, 1690

August brought a deterioration in the weather. Down in London, Evelyn noted that it broke on the 12th and up in the North it perhaps broke earlier: another expedition up Cross Fell with Messrs "Mayer",[80] Hume and Robinson seems to have been something of a fiasco – the walking was rough, it was cold and unpleasant and they saw nothing that Nicolson thought worth noting down except "Cotyledona Hirsutam". (*Saxifraga stellaris* L.).

Eleventh excursion: Shap, 11th September, 1690

The last excursion of the season was happier – with his old companions Hume and Lawson, Nicolson noted down the flora of the Shap area:

1. Alsine palustris Anglica. (*Sagina nodosa* (L.) Fenzl)
2. Androsaemum Hypericoides. (*Hypericum hirsutum* L.)
3. Carlina Syl*vestris. (Carlina vulgaris* L.)
4. Cornus foemina. (*Thelycrania sanguinea* (L.) Fourr.)
5. Equisetum nudum. (*Equisetum hyemale* L.)
6. Ferrum Equinum. (*Hippocrepis comosa* L.)
7. Hieracium λεπτόκαυλον.[81]
8. Lapathum Bononiense sinuatum.
9. Sedum Alpinum Luteum minus. (*Saxifraga aizoides* L.)
10. Sedum Alp*inum* trifido folio. (*Saxifraga hypnoides* L.)

This expedition seems to have ended the season's botanising and, with it, the first phase of Nicolson's emergent botanical interest. Sometime in this year of 1690, he opened his botanical notebook. As he ceased, thenceforward, to keep botanical records in his diary, one might fairly suppose that the notebook was begun in the winter when he could no longer get out into the field. A stray date in the notebook (page 47 below) shows him to have been busy tackling the difficult *hieracia* in the spring of 1691 and he may, again, have enjoyed summer expeditions in the company of Thomas Lawson. At the end of the year, however, the acquaintance came to an end with Lawson's death. On the 12th November, the best known of the early Lakeland botanists died and was buried in the Friends Burial Ground at Newby, near Great Strickland.

Nicolson, however, was by now a competent botanist. His expeditions of 1690 show him acquainting himself with the plants of different habitats in a workmanlike way. He observes the weeds of cultivated ground in and

---

80. "Mr Mayer", would be the Mayor of Carlisle, James Nicolson, a relative of William's. This was his usual style of address: See *CW2* ii 211.
81. fine-stemmed.

around his own parish (First expedition), he climbs Cross Fell and notices moorland plants (Fifth and Tenth expeditions) and, with Lawson, he looks at the plants of bogland (Sixth expedition). For rarities, he goes to gardens – both the gardens of the substantial gentry (Second expedition) and the gardens of such men as the discerning John Robinson (Third expedition) or of Thomas Lawson himself. From such groundwork, Nicolson built up a sound and thorough knowledge – the *Catalogue of Plants* speaks for itself.

Botanical Contributions to Published Works

In the early 1690s, the young Edmund Gibson of Bampton Grammar School, Cumberland, and lately down from Queen's College, Oxford, undertook the editing of Camden's *Britannia*, one of the most well-known achievements of the Elizabethan age and due, under Gibson's editorship, to become better-known than ever in his two handsome editions of 1695 and 1722.

At this time William Nicolson, who was some fourteen years Gibson's senior, was already noted as a scholar – he had left Oxford with a reputation as a leading Saxonist and it was known that, up in Cumberland, he was working on a history of the northern Counties of England.[82] This work was never completed, but some of the material that Nicolson collected for it went into the revision of Camden's section on Northumberland, which he was invited to undertake.

In addition to this, however, Nicolson produced a list of ten plants to supplement the very brief list that John Ray had made out for the County of Cumberland. On page 846 of the 1695 Camden appeared the following (numbers refer to the numbering of the list on pages 103–105):

> *An Additional account of some more rare Plants observ'd to grow in* Westmoreland *and* Cumberland, *by Mr.* Nicolson, *Arch-deacon of* Carlisle.

(5) Cannabis spuria fl. magno perelegante. *About* Blencarn, *in the parish of* Kirkland, *Cumberland.*

(9) Equisetum nudum variegatum minus. *In the meadows near Great* Salkeld; *and in most of the like sandy grounds in Cumberland.*

(*11*) Geranium Batrachoides longius radicatum, odoratum. *In* Mardale *and* Martindale, *Westm.*

(*14*) Hesperis Pannonica inodora. *On the banks of the Rivulets about* Dalehead *in Cumberland, and* Grassmire *im* [sic] *Westmoreland.*

(*16*) Orchis palmata palustris Dracontias. *Upon the old Mill-race at little* Salkeld, *and on* Langwathby-*Holm, Cumberland.*

(*17*) Cynosorchis militaris purpurea odorata. *On* Lance-moor *near* Newby, *and* on Thrimby-*Common, Westmoreland.*

(*24*) Serratula foliis ad summitatem usque indivisis. *Found first by* Reginald Harrison,[83] *a Quaker, in the Barony of* Kendal, *Westmoreland.*

---

82. See above page xxiv.
83. The Register of Quarterly Meetings of Westmorland Burials at Friends House, Euston Square, London, enters a Reginald Harrison as buried on 19th April, 1707 and possibly this is the man.

(*27*) Thlaspi minus Clusii. *On most Limestone pastures in both Counties.*

(*29*) Tragopogon Purpureum. *In the fields about* Carlisle *and* Rose-castle, *Cumberland*.

(*31*) Virga aurea latifolia serrata. C.B. *It grows as plentifully in our fields at* Salkeld *as the* Vulgaris; *which is as common as any Plant we have.*

The source of this list in the *Britannia* is to be found amongst the supplementary matter that Nicolson added into his notebook on the pages succeeding the *Catalogue of Plants* (pages 103–105 below). At some stage, he had access to Lawson's own copy of John Ray's *Catalogus Plantarum Angliae* – possibly he borrowed it from Lawson's daughter (Lawson being then dead) when he set out to compile the list for the *Britannia*. In the margin of Ray's *Catalogus* – or possibly on interleaved pages as some of the entries are rather long – Lawson had made notes about rarities and additional plants that he had observed. These Nicolson copied down on the spare sheets of his own botanical notebook and drew from them to produce the list for the *Britannia*. He was selective about this as some of the Lawson marginalia had already been published as they had been forwarded by Lawson himself in a letter to John Ray dated 9th April, 1688[84] and arrived in time to be incorporated into Ray's *Fasciculus Stirpium Britannicarum*. Amongst a long list of other plants, were the following from the marginalia:

Letter of Lawson to Ray 9th April, 1688	Nos. 6, 7, 8, 10, 13, 18, 19, 20, 22, 28, 31
Selected for publication in the *Fasciculus*	Nos. 6, 7, 8, 10, 20, 22, 28

When the *Synopsis Stirpium Britannicarum* came out in 1690, the Lawson records appearing in the *Fasciculus* were reprinted and Nicolson was careful to mark with a cross these and any other plants in the *Synopsis* of which Lawson also had a record in his marginalia. This added the following numbers to the list:

Nos. 2, 4, 13, 19, 25, 26, 30 and no. 29 where Nicolson quotes the *Synopsis* although he puts no cross against the plant.

From the remaining sixteen unpublished records, Nicolson made a selection of nine to send to Gibson for inclusion in the *Britannia*. He ignored plants which had previously been published and he ignored nos. 1 and 21 which are given in the marginalia with locations outside Westmorland or Cumberland and for which he himself had found no locations within these two counties. And he clearly had reservations about nos. 3, 12, 15 and 18 which are white flowered varieties of plants already well-known and no. 23 which is noted in the marginalia as a garden escape, although he failed to exclude no. 11 which Lawson thought might have been a garden escape, but

84. The text of Lawson's letter to Ray is to be found on pages 197–210 of Edwin Lankester ed., *The Correspondence of John Ray* (1848).

which Nicolson, finding afterwards in Mardale and Martindale, took to be native. To make up the ten, he included "Tragopogon Purpureum" (*Tragopogon porrifolius*), a plant which Ray, regarding as doubtfully native, had queried in the first edition of the *Synopsis*. Curiously enough, Nicolson received support for his opinion that it was a British plant from Dr. John Robinson or Fitz-Roberts who is quoted on page 77 of the second edition of the *Synopsis*, 1696.

Nor, in other respects, was Nicolson a mere copyist. At first sight, it may seem strange that the *Britannia* list should not be acknowledged to Lawson. In fact, comparison of this list with the original marginalia shows that Nicolson had done some field work himself on the records. In practically every case he added locations of his own or modified the Lawson location – usually by making it more general as his own observations made it clear that the plant was more widely distributed than had been thought. And in three instances (nos, 5, 14, 27), he gives Cumbrian locations where there had been none before.

In 1708, he was to return to the marginalia at the request of Thomas Robinson of Ousby. On 3rd September of that year his diary records:

> Mr Robinson left us and his papers; desireing a List of some rare plants: Which I promis'd to have in readiness agt his Return.[85]

The following day Nicolson, always prompt in the execution of business, records:

> Free from company, I wrote out a Sheet of rare plants for Mr. Robinson.[86]

An editorial footnote by Bishop Ware suggests that this was probably the list published in Robinson's *Essay towards a Natural History of Westmorland and Cumberland* (1709). It is, I think, undoubtedly so. The statement made by S. L. Petty that Robinson borrowed the *Catalogus Plantarum Angliae* with the marginalia in it from one of Lawson's daughters[87] would appear to be pure assumption based on Robinson's heading. The list appears on pages 89–95 of his work:

A LIST OF

> Several rare Plants, (not observed by Mr. *Ray*,) found in the Mountainous Parts of the Counties of Westmorland and Cumberland, by the late Eminent Botanist Mr. *Thomas Lawson*; and by him noted on the Margin of the said Mr. *Ray*'s Catalogue of *English* Plants, now in the Possession of (his Daughter) Mrs. *Thompson of Farmanby*.

---

85. *CW2* iv 38.
86. *CW2* iv 39.
87. S. L. Petty, "Thomas Lawson, the Father of Lakeland Botany", page 236, *The North Lonsdale Magazine and Furness Miscellany* (Ulverston), vol i, no. 12, pages 234–37.

(*1*) *Alsine Becaburgae* [sic] *folio*, Morisoni. *Chickweed*, with the Leaves of common *Brook-lime*; as described by Dr. *Morrison* of *Oxford*, very frequent.

(*2*) *Anonis Spinosa, flore Albo*. The prickly *Rest-harrow* (usually Purple) with a white Flower. On sandy Hillocks near the Sea-shore.

(*3*) *Armeria (sive Caryophyllus) pratensis, flore Albo*. White flower'd *Meadow-Pink*. This was observ'd near *Orton* in *Westmorland*; and growing in the like marshy Grounds, with the Common, cannot be suppos'd to be only a Variety, and not a distinct *Species*. The same perhaps may be said of the forementioned *Anonis*.

(*4*) *Bifolium Palustre, tribus foliis*. Three-leav'd *Tway-blade*. This was first found in the *Low-Hag* over against the Mill at *Great Strickland*. But he afterwards met with it in sundry other places of the Neighbourhood, as likewise elsewhere in the County of *Westmorland*.

(*5*) *Cannabis Spuria, flore albo magno & elegante. Wild Hemp*, with a large and beautiful white Flower. This was first observ'd, as an extraordinary Rarity, by Dr. *Merret*, in the Forest of *Sherwood*, and other parts of *Nottinghamshire*: It grows as plentifully on the Skirts of *Cross-Fell*, and other places within both these Counties.

(*6*) *Cardamine flore pleno*. Double flower'd *Lady's Smock*. This grows commonly enough on the Pasture Grounds near *Little Strickland*; and 'twas elsewhere found with no less than eight Rows of *Petali*.

(*7*) *Caryophyllata, flore amplo purpureo pleno*. Double flower'd *Herb-Bennet*, with a large and purple Flower. This comely Plant was sent by the Discoverer (Mr. *Lawson*) to Mr. Ray; who acknowledged it to be hitherto undescrib'd, and therefore bestow'd upon it the following Description, *Caryophyllata purpurea prolifera, quadruplici aut Quintuplici serie petalorum; e medio floris emergit caulis, florem in summitate gerens octodecim petalorum.*

(*8*) *Cotula non foetida, flore pleno*. Double flower'd *Dog's Camomile*. This Rarity, with four or five Ranks of Flowers, was met with in the Discoverer's own Grounds at *Great Strickland*.

(*9*) *Equisetum Nudum variegatum minus*. J.B. the naked and party colour'd *Horse-Tail* of *John Bauhinus*. This was first shew'd to *Mr. Lawson* at *Great Salkeld*, but grows in so great plenty there, and every where on the Banks of the River *Eden*, that he could not but wonder that this was the first time of its being observ'd in *England*; 'tis an early, and quickly fading *Vernal Plant*, which might probably be the occasion of its not being hitherto taken notice of by those Curious Gentlemen, who commonly began their Circuits too late in the Year for such a Discovery.

(*10*) *Geranium Batrachoides flore eleganter variegato. Crowfoot, Cranes-bill*, with a beautiful party colour'd Flower. A dry'd Sample of this, found in Mr. *Howard*'s Park at *Thornthwait*, was sent to Mr. *Ray*, who (in his Supplemental *Fasciculus* soon after publish'd) took notice of it as a special Rarity.

(*12*) *Geranium Columbinum folio malvae rotundo, flore albo. Doves-foot, Cranes-bill*, with a white Flower. This being observ'd several

Years together in good fruitful Ground, under a Wall near the *Round-Table* at *Eamont-Bridge*, the Discoverer thought he had reason to reckon as a new *Species*; tho' he doubted whether he might boldly say the same of that which follows.

(*13*) *Geranium Haematodes album, venis rubentibus striatum*. Bloody *Cranes-bill*, striped with redish Veins. The Mixtures in the Common kind (tho' even that is peculiar to those and the like *Alpine* Countries) are quite contrary. This *Variety*, for he suppos'd it to be no more, was found on several sandy Grounds near *Millum* in *Cumberland*, but most plentifully in the Isle of *Walney*.

(*15*) *Narcissus Flore albo & albido*. The Common wild *Daffodil*, with a white and pale colour'd Flower. The latter of these is frequently observ'd to grow intermix'd with the Ordinary yellow; but the former was first gathered near his own House at *Strickland*, and afterwards near *Ulverston* in *Lancashire*.

(*17*) *Orchis Militaris purpurea odorata*. *Parkinson*'s sweet purple flower'd Soldier's *Cullions*. This was look'd upon as a choice Rarety, when he first met with it, (about the *Fairy Holes*) on *Lancemoor* near *Newby* in *Westmorland*: But 'twas afterwards found abundantly in the Meadows upon both the Banks of *Eden*, throughout several Parishes.

(*19*) *Pedicularis Palustris elatior alba*. The larger Meadow *Lowse-wort,* or *Rattle*, with a white Flower. This grows pretty plentifully near the Foot of *Long Sleddal*, by the side of the common Road, leading towards *Kendale*.

(*20*) *Ptarmica Flore pleno*. Double flower'd *Sneezewort*. In one of the little Islands, call'd *Small Holme*, in the great *Lake* of *Winander-Meer*.

(*23*) *Scabiosa montana maxima Lobetii* [sic]. The great Mountain *Scabious*. This Plant is well known to be a Native of the *Italian* and *Helvetick* Alps; and Mr. *Lawson* reasonably enough concluded from thence, that it might also have a spontaneous Growth in this Country, when he found it near the Lord *Lonsdale*'s Seat at *Lowther*; but he was afterwards rather inclined to believe (as he confesses) that the place where he gathered it, had probably been heretofore a Garden.

(*25*) *Thlaspi Veronicae folio Parkinsoni*. This is the same sort of *Penny-Cress*, which *John Bauhinus* calls by the Name of *Bursa pastoris loculo sublongo affinis pulchra planta*. It grows on the moist sides of many of our Northern Mountains. Its seminal Leaves (which lie next the Ground) are rough, hairy, almost round, indented, of a deep green Colour, each on a short foot Stalk, somewhat resembling the Leaves of *Speedwell*; and its Stalk is also hairy, half a Foot high, branching usually from the bottom, though sometime without Branches. At the Top are many small white Flowers, which are succeeded by small long Pods, one above another, Spike-fashion. In each of these there is a slender brownish Seed. The Root is very white and long.

The omission of all mention of Nicolson in this list might well have been at his own request. He had already been an unwilling participant in the controversy that Robinson had raised over his theories of the Origin of the Earth,[88] and he had probably now no wish to be further involved with Parson Robinson's enterprises. The diary entries, the fact that Nicolson had his own copy of Lawson's marginalia and the fact that Nicolson had achieved some notice as a botanist and was fit to comment on these (which Robinson, probably, was not) all seem to admit of no doubt that the list printed in Robinson's *Natural History* originated with Nicolson. By 1708, however, Nicolson's episcopal duties left him little time to pursue his interest in botany and whatever enthusiasm he had for Natural History went into collecting fossils and stones. So the residue of Lawson's marginalia came out attributed to their author to whom they most properly belonged. A few later found their way into the 1724, Dillenian edition of Ray's *Synopsis*, and are quoted with reference to Thomas Robinson. Below is the history of the publication of this second group:

Selected by Nicolson and sent to Thomas Robinson for publication in *The Natural History*, 1709	1, 2, 3, 4, 5, 6, 7, 8, 9, 10, 12, 13, 15, 17, 19, 20, 23, 25
Published in the Dillenian edition of Ray's *Synopsis*, 1724	5, 8, 9, 12, 13, 15, 17, 19, 20

The Dillenian edition was also to include some of Nicolson's *Britannia* records which had originated with the marginalia: namely nos. 14, 16, 27, 29 and 31 which are referred to him, being the only references to Nicolson in the *Synopsis* apart from his record of the form of *Potentilla palustris* with leaves densely villous beneath ("Pentaphyllum palustre rubrum crassis et villosis foliis") which was received by Doody and appeared in *Synopsis*, 1696 (Appendix), page 326 of Vol 3 of the *Historia Plantarum*, 1704, and again in *Synopsis*, 1724 (page 256).

Thus it will be seen how Nicolson used the data at his disposal: respecting Lawson's records, he added his own locations with critical comment that displays independent judgment and scientific caution, realising that plants once thought rarities may be commoner than supposed and being careful in matters of duplicated species and garden escapes.

Robinson's *Natural History* was, however, not quite the end of the records. The *Britannia* list was to reappear in the Cumberland section of Thomas Cox's *Magna Britannia*, lifted direct from Camden with the omission of two plants (nos. 11 and 24) which have Westmorland locations, and with the addition of a short list of five plants compiled by Ray that precedes the Nicolson list in Camden and which Cox takes as being found in Cumberland, failing to notice the prefixes for Lancashire and Westmorland before two of them: "Lan. Eruca Monensis laciniata lutea" and "W. Gladiolus lacustris Dortmanni". Preceding the whole is a note, again adopted from Camden:

---

88. See above page xxiv.

Besides the Herbs and Plants which are common to this County with others, and such as may be esteemed peculiar to it of *English* Growth, there is a Tradition among the Inhabitants who dwell near the Wall or *Vallum*, That the *Roman* Garrisons upon the Frontiers planted in the Country about them an abundance of Medicinal Plants for their own Use brought from other Countries, and it is so firmly believed, that the *Scotch* Empiricks, and many other Persons who are curious, come hither out of *Scotland* on purpose every Year in the beginning of the Summer to gather them, affirming, that they, by long Experience, have found them to be of a much more sovereign Virtue than those in other Places, and especially such as they find nearest the Wall: But if the *Scotch* Physicians and Surgeons do still keep up such a Custom, 'tis only intended to deceive the People; for the Rt. Rev. the Bishop of *Carlile* (then Mr. *Nicholson*) after the most diligent Search, could not meet with any sort of Plants growing along the Wall, which are not as plentiful in some other Parts of the Countrey.[89]

Seemingly nothing, however, could displace the popularity of the original *Britannia* and, in 1722, a second Gibson edition was published. In this was reprinted Nicolson's original list from the 1695 edition, together with a note that he was now Lord Bishop of Derry and the list was retained in the subsequent reprints of Gibson in 1753 and 1772. In 1789, however, just over 200 years after the appearance of the original squat little Elizabethan volume, Richard Gough undertook yet another edition of Camden and, with it, a full revision of the plant lists under Linnaean binomials. From this revision, only three of Nicolson's list emerge unscathed to make their final appearance in a reprint of the Gough edition in 1806:

(*14*) (Gough, page 418, Westmorland): Hesperis *inodora*. Unsavory Dame's Violets; on the banks of the rivers about *Dalehead* and *Gresmere*.

(*27*) (Gough, page 420, Westmorland): Thlaspi *alpestre*. Perfoliate Bastard Cress; on the moist (sic) limestone pastures.
(page 463, Cumberland): ... In the moist pastures in a limestone soil.

(*29*) (Gough, page 463, Cumberland): Tragopogon *porrifolium*. Purple Goat's-beard; in the fields about *Carlisle* and *Rose* castle.

With this may be said to be ended the continuous history of the publication of the Nicolson records. The notebook itself went to Ireland with the Bishop and disappeared from sight until the Cotton sale of 1868 whence it returned to its native Cumbria. There it was to come to the attention of William Hodgson when he was compiling his *Flora of Cumberland* (1898). In the preface to that work, he acknowledges his indebtedness to the then Bishop of Carlisle, Dr. Hervey Goodwin, in allowing him to make extracts from the manuscript for the purposes of his own flora, a study of which

89. Thomas Cox, *Magna Britannia* (published in parts 1720–31), vol i, page 398.

reveals that he selected some 86 plants for attention. Some of these received more detailed comment in a paper entitled "Disappearance of Plants in Cumberland", which was published in *The Naturalist*[90] in January, 1891.

The latest publication of any Nicolson records that I can find is in the Victoria County History.[91] In his introduction to the botanical section, Hodgson again cites the MS *Catalogue of Plants* and in the subsequent botanical list there are several references to it.

90. Pages 7–12.
91. *V.C.H.*, 1905, vol ii, Cumberland, page 73 ff. It is to be noted that Hodgson generously bestows upon Nicolson the Parish of Great Strickland.

Nicolson as a Botanist

In 1690, Nicolson sat down and made the first entry in his botanical notebook: "Abies. Fir tree". Like John Ray, when he compiled the *Catalogus Plantarum Angliae*, Nicolson chose a simple alphabetic arrangement for his own catalogue. And, modelling it upon Ray's, he chose a similar title: *Catalogus Plantarum Britanniae*. To this finding list of British plants, he proposed to add locations within the ancient Kingdom of Northumbria – an area defined by Camden as comprising the counties of Lancaster, York, Durham, Cumberland, Westmorland, Northumberland and Scotland as far as the Firth of Forth.[92]

With this plan in mind, Nicolson set about the compilation of the finding list. The Raian records available to him, other than the first two volumes of the monumental *Historia Plantarum* (1686 and 1688), were the two editions of the *Catalogus Plantarum Angliae* (1670 and 1677), together with its supplementary *Fasciculus* of 1688 and – if we assume that Nicolson did not begin the catalogue until late in 1690 – the *Synopsis Stirpium Britannicarum*.

Examination of the manuscript establishes that the basis of his list is the *Catalogus Angliae* in its second edition, combined with the *Fasciculus*. In a preface to the second edition of the *Catalogus Angliae* Ray lists the plants new to this edition. All but a handful are listed in Nicolson's *Catalogus Plantarum Britanniae* and none show signs of having been squeezed in as late entries. Moreover, the considerable additions that Ray made to the fungi in the 1677 edition of the *Catalogus Angliae* are reflected in Nicolson's own catalogue.

This evidence is plain enough and, equally plain, is the part played by the *Fasciculus* in the first compilation of the catalogue. There are nearly 150 plants that make up the *Fasciculus* list and Nicolson incorporates practically all of these into his own list with little evidence of late insertions. There is, of course, the occasional doubtful case, but Nicolson was working from two sources and it is possible that, in such cases, the plant was squeezed in simply because he got into trouble with the alphabetic order. The bulk of the entries show a congruity with the *Catalogus Angliae* entries which is only explicable in terms of their having been written down all at the same time.

Lastly, we have to consider the *Synopsis Stirpium Britannicarum* which appeared in May, 1690. On the 17th of that month, we find the London-based Sherard promising to send a copy up to Richard Richardson in Yorkshire.[93] It seems, however, that Nicolson either compiled his finding

92. Edmund Gibson ed., *Camden's Britannia Newly Translated into English* (1695), col clxvi.
93. Dawson Turner ed., *Extracts from the Literary and Scientific Correspondence of Richard Richardson* (Yarmouth, 1835), page 2.

list before that date,[94] or did not obtain his copy until much later. His use of the *Synopsis* is extensive and methodical, but it quite clearly post-dates the original compilation. There is evidence of squashed entries, duplicate numbering and additional cross-referencing of species. Amongst the *Hieracia*, for instance, there are three squeezed entries: "Hieracium echiodes capitulis cardui Benedicti", "Hieracium leptocaulon hirsutum folio rutundiori" and "Hieracium montanum augustifolium" – duplicating numbers 3, 4 and 10. Elsewhere Nicolson has inserted cross-references that are not in the *Catalogus Angliae*, either because a study of the *Synopsis* revealed these to him or because he himself was trying to make his work more comprehensive. He does not, in any case, stick to the very letter of John Ray's work, sometimes preferring one synonym to another or harking back to the records of older botanists such as Gerard or Johnson, and he consults the work of other contemporaries such as Robert Plott, who included plant lists in his histories of Oxfordshire and Staffordshire.

The publication of the *Synopsis Stirpium Britannicarum* was to prove a milestone in the history of botany – it was a work that was not superseded for three quarters of a century. A second edition was published in 1696 and the much enlarged edition of J. J. Dillenius in 1724 – by which time Nicolson had moved to Ireland and was nearing the end of his life. Long before that his interest in the *Catalogus Plantarum Britanniae* had become spasmodic. In the *Synopsis* of 1696, is recorded the notable find of his friend Edward Lhwyd "Bulbosa Alpina juncifolia ..."[95] or *Lloydia serotina*, the plant which commemorates his name. It was not added by Nicolson to his finding list, although he might have been expected to think that he might find it in Cumbria.

Between Nicolson and Lhwyd, there is a curious similarity in the progress of their interests. They were both keen philologists and antiquarians and both, starting as keen botanists, were diverted by fossils and by geological speculation. Both, moreover, were interested in the mountain flora of Britain and it is faintly disappointing that Nicolson found no leisure, or perhaps had no inclination, to visit the high fells of central Lakeland. His 175 pages of clear and careful manuscript are, after all, only the part-time records of a man who was not only an energetic and ambitious clergyman, but also a scholar in other branches of learning in which he had greater repute. His contemporaries, however, thought well of him as a botanist and he contributed records to Camden's *Britannia*. But he lives in the shadow of the greatest of all British botanists, John Ray. He has recorded for us many Cumbrian plants – but it was the Quaker botanist, Thomas Lawson, his senior by 25 years, who earned himself the title of "Father of Lakeland Botany". And though some notice of his botanising efforts has filtered through to us by way of such works as Hodgson's *Flora* and the Victoria County History, there has until now been no edition of his notebook. He aimed to compile a flora of the counties that had once comprised the ancient

94. I think this unlikely. See above page xxxvii.
95. Page 233.

Kingdom of Northumbria, but he never published it and his records for plants outside Cumberland, Westmorland and Furness are negligible and, for the most part, not his own. Within these counties, however, his activity is impressive as a glance at the map and statistical information on pages 124–25 below will show and it is as characteristic of the man as any other piece of scholarship on which he was engaged. William Nicolson was, above all things, a compiler of records. One has only to look at other aspects of his many-sided scholastic achievement to see this. In his youth, he was something of a lexicographer, for he edited Junius' Anglo-Saxon dictionary and, if we jump the span of his years, we find that his last published work was the *Irish Historical Library*, 1724, a work that completed his bibliographic survey of the resources available to the historian – a programme that had begun with *The English Historical Library*, 1698–99 and *The Scottish Historical Library*, 1702.

All his life Nicolson enjoyed method and order – it could not have been so full a life had it been otherwise. He was not, as much through temperament as lack of leisure, a typical virtuoso of the late seventeenth century. On his trips to London after his promotion to the see of Carlisle, he enjoyed talking the latest scientific shop with the experimenters and virtuosi of the Royal Society. But neither the circumstances of his life nor the fact of his geographical isolation in Cumbria were conducive to the experimental philosophy. He made extracts from Nehemiah Grew's *Musaeum Regalis Societatis* (see below page 106), but he gave no sign of having been interested in the work that Grew did on plant anatomy and chemistry. No machinery for investigating aspects of plant physiology was ever, so far as we know, set up in the library at Rose Castle and his diary does not reveal that he ever owned a microscope. The pleasures of collection, rather than speculation, and a strong vein of local patriotism were the background to the *Catalogue of Plants* and, with it, Nicolson joins the great company of clergyman naturalists, two of whom, William Paley (1743–1805) author of *Natural Theology*, and the late Canon Hervey, founder of the Lake District Naturalists Trust, were to be successors to Nicolson in the Rectory at Great Salkeld.

In the years that he was at Great Salkeld, 1946–67, Canon Hervey himself made records of the flora of the parish and his manuscript notebooks are now kept at Brathay Hall Field Study Centre, near Ambleside. Comparison of these records with Nicolson's show considerable changes in the vegetation of the parish. Both men record something approaching 400 plants from the area, although it is not easy to make exact comparisons between two botanists who are separated by a gap of 250 years and whose terminology lies either side of the great divide of Linnaeus. Only just over half of these, however, are common to both. Included in the remainder are two groups which one might wish to ignore in order to compare like with like. On Nicolson's part, there are over 40 non-flowering plants which need to be excluded from consideration and one might, similarly, wish to exclude a group, roughly equal in number, of reeds, sedges and rushes which were noticed by Canon Hervey. It is true that Nicolson aimed to include these in his records and indeed takes notice of a few. But his records are so imperfect and the subject was so much in its infancy in his day that one might, for the purposes of comparison, set aside the few records that he does have.

Of the remaining plants found by Canon Hervey at Great Salkeld, not all were unknown to Nicolson. Just over 40 were recorded by him from other locations within Cumbria and some from nearby parishes – the Common Bog Asphodel, *Narthecium ossifragum*, for instance, is recorded from Cliburn Moss and Renwick Holms and one might, perhaps, consider that if he describes the Yellow Pimpernel, *Lysimachia nemorum* as growing "in bogs common", then he may well have seen it within the bounds of his own parish.

Of the plants left, an enormous proportion are, as one might expect, aliens and stragglers unknown in Nicolson's day. *Veronica persica*, known in Britain for only about 200 years, and *Veronica filiformis* might be said to come into these categories, as might the nineteenth century introductions *Mimulus guttata*, the Monkey Flower and the Himalayan Balsam, *Impatiens glandulifera*, so conspicuous by riversides. And there are several plants belonging to difficult groups – some not even distinguished as separate species in Nicolson's day – which Canon Hervey, profitting by two and a half more centuries of taxonomic study, was enabled to identify. Nicolson is conspicuously weak on *Ranunculaceae*, for example, and amongst his mouse-ears, sandworts and pearlworts there are several omissions that are perhaps ascribable to this cause.

For his own part, in cataloguing the Salkeld of the late seventeenth century, he lists plants and weeds that belong to a pattern of agriculture that had changed radically by Hervey's day. He notes the crop plants, horse beans and buckwheat, hops and flax and hemp, and with them weeds that are uncommon in the fields of today. His parishioners would have seen, though perhaps not enjoyed, the yellow Charlock, *Sinapis arvensis*, the Corn Cockle, *Agrostemma githago* and the Corn Marigold, *Chrysanthemum segetum* and perhaps they made efforts to uproot the poisonous Henbane, *Hyocyamus niger*, and the Woody Nightshade, *Solanum dulcamara*, which has suffered for its name. More pleasing to the eye, perhaps, would have been the Throatwort, *Campanula trachelium*, and the Mealy Primrose, *Primula farinosa*, a plant of a wetter Salkeld than Hervey's. And perhaps Nicolson's parishioners sometimes had resort to medicinal herbs such as Betony, *Betonica officinalis*, to which Nicolson gives the Old English name Attor-lathe, meaning a counter-poison. Other medicinal herbs which grew in the Salkeld of his day were the two Horehounds, *Ballota nigra* and *Marrubium vulgare*, and the three common Woundworts, *Stachys sylvatica, S. arvensis* and *S. palustris*, which was used by Andrew Marvell's mower to staunch the wound made by his own scythe in the song "Damon the Mower".

One might perhaps have expected more water-loving plants than Nicolson lists. The Cumbria of his day contained many more areas of marsh and bog and he himself loved to botanise down in the water meadows of his own and neighbouring parishes – the "holms" to which he refers so often. And one might, in an age when everyone was discovering novelties, have expected him to have hazarded a few novae of his own. Thomas Lawson was quick to observe variations and to tack "flore pleno" or "flore albo" to known plants, but Nicolson was more cautious – a good deal of the pioneer work had been done by his day and the list of Ray's *Synopsis* was a very full list to work upon. And he was not, one must say, an innovator. We see him, on page 57 below, trying to grapple with the *Epilobiums* and, as we have seen, he brought his list for the *Hieracia* up-to-date after the publication of the *Synopsis*. But he shows little other signs of exploration of difficult groups, although he was obviously quite interested in ferns and the manuscript shows several additions to the sections listing these. He made, however, no discoveries and even where he had hoped to make discoveries – in the mountains of Cumbria – he has not given us anything new. The "new faces" that he writes to tell Lhwyd about from time to time are new to him, not new to science, although very often they are first Cumbrian records. His fell or moorland sites are very restricted, most of this sort of botanising being done in the Shap area with the occasional expedition up Cross Fell. And occasionally, too, his duties took him over the Alston road to Durham or Newcastle. From Dumma Hill on Stainmore, he records the pretty Mountain Pansy, *Viola lutea*, where, as long ago as 1685, he had noticed "Flos trinitarius wild on ye marshes near Bowes, and of several colours. Dumme Hill near Burgh"[96] (*Viola tricolour* and probably *Viola lutea* also). Entries such as this do not always find a place in his notebook – he climbed Blencathra in 1704 and saw *Lobelia dortmanna* in Bowscale tarn, but it is not added to the locations already given in the *Catalogue of Plants*. And a letter to Lhwyd of 22nd June, 1693 shows how he shared Lhwyd's special interest in mountain flora:

> And now we talk of plants, it will not, perhaps, be unacceptable to tell you, that I took the pains to climb a deal of our hills in quest of alpine plants. I found vast quantities of several mentioned by Mr. Ray from Mr. Lawson as great rarities, and confined to some particular places: for example, the *Acetosa rotundifolia, Rhodia Radix*, and *Alchymilla Alpina*, I met with on all the rocks near the tops of the ragged mountains, for twenty miles together: and, I believe, few of the like kind want them.[97]

Such communications show him extending the range of his botanising in a way that, perhaps, the catalogue does not. And there are other omissions in the *Catalogue* that one can only attribute to scribal error – he gives no record of having seen the meadow buttercup in Cumbria. If he had intended to publish, he would no doubt have tidied these things, and one should bear in mind, when looking at the work, that he never proceeded with it beyond the stage of a notebook. Had he made of his records of Cumberland, Westmorland and Furness, a volume complete in itself it would have been the first regional flora to succeed the great John Ray's *Cambridge Catalogue* of 1660.

---

96. *CW2* i 30.
97. Ed Nichols, *Letters ... to and from William Nicolson,* vol i, page 36.

AUTHORITIES AND REFERENCES

There follows a list of authorities and references which occur in the text and need to be glossed. In basing the *Catalogue of Plants* very largely upon the works of John Ray, Nicolson adopted many of the abbreviations used by Ray. A description of the works signified by these abbreviations will be found on pages 82–8 of William T. Stearn's edition of Ray's *Synopsis Methodica Stirpium Britannicarum, 1724* (Ray Soc., 1973), where references to the appropriate entries in G. A. Pritzel, *Thesaurus Literaturae botanicae,* 2nd ed. (1872), are also given. References below are to Agnes ARBER, *Herbals, their Origin and Evolution,* 2nd ed. (Cambridge, 1938); Ray DESMOND, *Dictionary of British and Irish Botanists and Horticulturists* (1977); and Geoffrey KEYNES, *John Ray, a Bibliography* (1951).

AS — Anglo-Saxon.

Aelf. Gloss. — Aelfric's Glossary. This had been printed by William Somner at the end of his *Dictionarium Saxonico-Latino-Anglicum* (Oxford, 1659).

B.P. — J. P. Cornut, *Canadensium Plantarum aliarumque nondum editarum Historia. Cui adjectum est calcem Enchiridion botanicum Parisiense* (Paris, 1635).

C.B. — Caspar or Gaspard Bauhin. b. 1560. Professor of anatomy and botany, Basle, and later Professor of medicine. d. Basle 1624 (ARBER 114–16).

Cam. — Joachim Liebhard, known as Camerarius (or Kammermeister). b. 1534, Nuremberg. Gardener. Editor of Matthioli. d. 1598 (ARBER 76, 77, 78).

Clus. Clusij — Charles de l'Ecluse. b. Arras 1526 and travelled over Europe botanising. Author of *Rariorum Plantarum Historia* (Antwerp, 1601). d. Leiden 1609 (ARBER 84–9).

Diosc. — Dioscorides. Greek botanist who fl. 1st century A.D. Author of the *Materia Medica* (ARBER 8–9).

Matthew Dodesworth	Rev. Matthew Dodsworth. b. Badsworth in Yorkshire 1654. Rector of Sessay from 1690 to 1697. Particularly interested in ferns. d. Sessay, Yorkshire in 1697 (DESMOND 189).
Dodonaei	Rembert Dodoens. Belgian botanist b. Mechelen 1517. d. Leiden 1585. Author of *Crüÿdeboeck*, Antwerp 1554 which was translated into French by Charles de l'Ecluse, appearing as *Histoire des Plantes*, Antwerp 1557. In 1578, this French translation was made known to English readers under the title *A Nievve Herball* by Henry Lyte (ARBER 82, 124–25).
fasc., Fasc., Fasc. Stirp.	John Ray, *Fasciculus Stirpium Britannicarum* (1688). (KEYNES no. 9). See also J. R.
Ger.	John Gerard, *The Herball or Generall Historie of Plantes* (1597). b. Nantwich, Cheshire 1545. Barber-surgeon. Garden in Holborn, London, of which he published a catalogue in 1596. d. London 1612 (DESMOND 248).
Ger. Em., Ger. emac.	John Gerard, *The Herball or Generall Historie of Plantes very much enlarged and amended by Thomas Johnson* (1633, 1636). See also Johnson.
Gesneri	Konrad Gesner. b. Zurich 1516. First Professor of Philosophy at Zurich, subsequently Professor of Natural History. d. Zurich 1565 (ARBER 110–13).
I.B., J.B.	Johann or Jean Bauhin. b. Basle 1541. d. Mumpelgard 1613. Brother of Caspar Bauhin (q.v.). Physician and botanist. Friend of Gesner (q.v.). (ARBER 113–114)
Johns. Johnson Johnston	Thomas Johnson, *Mercurius Botanicus* (1634–41). b. Selby, Yorkshire c. 1597. Apothecary. Had physic garden on Snow Hill, 1633 and was killed fighting for the king in the Civil War, d. of wounds, Basing House, Hants, Sept., 1644. Revised Gerard's *Herball* (DESMOND 346–7).

J.F.R.	John Robinson, alias John Fitz-Roberts. Botanist of Kendal where Nicolson visited his garden (see above pages xxxii–xxxiii) fl. 1690s–1710s (DESMOND 525).
J.R. Joh. Raij	John Ray. b. Black Notley, Essex 1627. Leading seventeenth-century British naturalist. Life by Charles E. Raven (Cambridge, 1942). d. Black Notley 1705. For works referred to by Nicolson, see Fasc. above and Raij below (DESMOND 513).
Doctor Lloyd	Edward Lhwyd, Lhuyd or Lloyd. b. Flanfihangel, Cardigan, 1660. Geologist, botanist, philologist, antiquary and keeper of the Ashmolean Museum 1690–1709. Friend and correspondent of Nicolson. d. Oxford 1709 (DESMOND 385).
Lob. Lobelij	Matthias de Lobel. b. Lisle 1538. Settled in London where he became superintendent of the garden of Lord Zouch at Hackney. d. 1616.
Matthioli	Pierandrea Matthioli. Physician and commentator on Dioscorides. b. Siena 1501, d. 1577 (ARBER 92–7).
Merret, Mer.	Christopher Merret or Merrett. b. Winchcombe, Gloucestershire, 1614. Glass maker. First Keeper of the library and museum of the College of Physicians, 1654. d. London 1695. Author of *Pinax Rerum Naturalium Britannicarum* (1666). (DESMOND 435).
Meth. Syn.	See Raij Synops. below.
Morisoni, morisoni	Robert Morison. b. Dundee 1620. Took doctorate at Angers, 1648 and returned to Britain after the Restoration. First Professor of Botany, Oxford 1669. d. London 1683. (DESMOND 450).
Muntingij	Abraham Munting. b. Groningen 1626. d. Groningen 1683. Author of *De vera antiquorum herba Britannica* (Amsterdam 1681).
Newtoni	James Newton. b. 1639. M.D. Friend of John Ray. d. 1718 (DESMOND 463).

nostras, nostratibus	our, with us (sometimes used in the sense "locally")
Officinarum	Derived from "opificina" or "officina", a workshop which came, in the middle ages, to be used of an apothecary's shop. The equivalent "officinalis" is still used as a specific term and might be rendered "medicinal".
Park. Parkins.	John Parkinson, *Theatrum Botanicum* (1640). b. Notts 1567. Apothecary and gardener and kept a garden in Long Acre, London. d. 1650.
Pl. Oxf., Pl. Staff. Doctor Plott	Robert Plott. b. Sutton Baron, Borden, Kent 1640. d. Sutton Baron 1696. Author of *Natural History of Oxfordshire* (1677), and *Natural History of Staffordshire* (1688). First Keeper of the Ashmolean Museum (DESMOND 498).
Plinij	Pliny the Elder, Gaius Plinius Secundus. Roman naturalist. b. A.D. 23 or 24, d. 24th Aug. 79 when observing the eruption of Vesuvius. Author of *Naturalis Historia,* translated by Philemon Holland (1601).
Raij, Mr Ray	John Ray. See above J.R.
Raij Catal. Exot.	John Ray, *Catalogus Stirpium in Exteris Regionibus A nobis Observatarum* (1673) published in John Ray, *Observations ... made in a Journey* (1673), (KEYNES no. 21).
Raij Hist., Raij Hist Plant	John Ray, *Historia Plantarum* (1686–1704), (KEYNES nos. 48, 49, 51).
Raij Synops., R. Synops Mr. Ray's Synopsis	John Ray, *Synopsis methodica Stirpium Britannicarum* (1690), (KEYNES no. 54).
Sch. Bot. par.	Warton, S. (Sherard, W.) *Schola botanica* (Amsterdam, 1689).
Sibbald Sir R. Sibbald	Sir Robert Sibbald. b. Edinburgh 1641. First Professor of Medicine, Edinburgh, 1685. Together with Andrew Balfour, started first botanic garden at Edinburgh. Author of *Scotia Illustrata* (1684). Correspondent and acquaintance of Nicolson. d. Edinburgh 1722 (DESMOND 558).

Sutherl. Ja. Sutherland	James Sutherland. b. c. 1639. Professor of Botany, Edinburgh 1695–1706. Published a catalogue of the Edinburgh botanic garden, *Hortus Medicus Edinburgensis* (1683). d. Edinburgh 1719 (DESMOND 592–3).
Synop. Synopsi(s)	See Raij Synops. above.
T. L., Th. Lawson Tho. Lawson Mr Lawson	Thomas Lawson. b. near Settle, Yorkshire 1630. Correspondent of Ray. Accompanied Nicolson on botanising expeditions in 1690 (see pages xxxv–xxxvii). Schoolmaster at Great Strickland where he d. 1691 (DESMOND 376).
Tabern.	Jacob Dietrich or Jacobus Theodorus of Bergzabern, therefore known as Tabernmontanus. b. c. 1520 and d. 1590. Author of *Neuw Kreuterbuch,* (Frankfurt, 1588–91) (ARBER 76).
Theophrastus Theophr. Theoph. Theophrasti	Greek botanist. b. 370 B.C. at Eresos in Lesbos. Pupil of Plato and Aristotle. d. c. 285 B.C. Author of *History of Plants.*
Tournefortij Tournf.	Joseph Pitton de Tournefort. b. 1656 at Aix in Provence. Professor at Jardin du Roï 1683. Travelled widely in Europe, Asia and Africa. d. 1708.

NOTE ON THE TEXT

Contracted Latin terminations have been silently expanded, also:

Ab^{t}. ab^{t}	=	about
ag^{t}	=	against
&	=	and *or* et
&c	=	etc
f.	=	forte
fl.	=	flore
fol.	=	folio
frō	=	from
G. Gr. G^{t} g^{t}.	=	Great
i.	=	i.e.
It.	=	Item
L.	=	Little
Nostr. nostr.	=	nostras
o^{r}	=	our
V.	=	Vide
ł	=	vel
w^{ch}	=	which
wth wthin wthout	=	with, within, without
y^{n}	=	than
y^{t}	=	that

The abbreviation Q., Qu., or qu. has not been expanded to *quaere* from which the Anglicised "query" derives. It has been treated, rather, as a piece of punctuation, analagous to "?", its modern descendant.

Greek ligatures have been silently expanded.

The abbreviation sylv. has been expanded sometimes to sylvestre or sylvestris, sometimes to sylvaticus – a – um, following the usage of the 1724 edition of John Ray's *Synopsis Methodica Stirpium Britannicarum*, which has been used for general guidance on the expansion of polynomial terms.

It has not always been possible to distinguish Nicolson's capital letters from his small: the letters m, n, c, o, p, s, u, v, w all present difficulties and I have used my discretion over these.

Square brackets have been used for deletions, angle brackets to enclose anything for which there is no manuscript authority. Nicolson himself used braces to emphasise the divisions between various groups of plants. In print, however, these divisions are quite clear and the braces have, accordingly, been omitted. \ / enclose insertions.

In the manuscript, Nicolson's general practice is to underline English names of plants in plummet, place names in ink. The two types of underlining have been represented in the text as follows:

Myosuros. **Mouse-tail.** CARTMEL-Grainge.

Elsewhere underlining has been represented by italics.

Quae notantur **S** in Campis, pratis et pascuis, infra parochiam de SALKELD sponte nascuntur.

Those plants which have an **S** *marked against them, grow wild in the open lands, meadows and pastures in the parish of Great Salkeld.*

CATALOGUS PLANTARUM BRITANNIAE

AD FIDEM CLL.

Johannis Ray,
Thomae Lawson,
Rob^ti^ Plott,
Aliorumque.

Cum Synonymis Graecis, Gallis, Germanis, A. Saxonicis, etc. Ad naturam cujusque explicandam inservientibus. Additis insuper Locis singularum Nativis per Comitatus *Cumbriae* et *Westmeriae*, caeterasque Regni veteris *Northymbrici* Ditiones.

A CATALOGUE OF BRITISH PLANTS

BASED ON THE WORK OF SUCH DISTINGUISHED SCHOLARS AS

John Ray
Thomas Lawson
Robert Plott
and Others,

with botanical names added from Greek, French, German, Anglo-Saxon, etc., to throw light on the various kinds of plant described, and with locations given for the various plants in the counties of Cumberland and Westmorland, as well as the other regions of the old kingdom of Northumbria.

Magnus noster in re Botanica, Coryphaeus, Johannes Ray, Totam Plantarum Brittanicarum Supellectilem in 26 Classes (seu Genera Summa) Divisit. Quae sequuntur.[98]

I. *Plantae Imperfectae.* haeque;
 1. *Submarinae,*
  1. *Lapideae.* Corallium album. Corallina, Alba et Rubra.
  2. *Corneae.* Una Species D*octor* Newton
  3. *Herbaceae.* Fuci et Musci.
 2. *Fungi.* qui –
  1. *Terrestres*
   1. *Lamellati.*
   2. *Lamellis carentes.*
  2. *Arborei.*
 3. *Musci.* qui –
  1. *Terrestres*; et Lichenes steriles.
  2. *Arborei.*
  3. *Aquatici.* Conferva. Lenticula.

II. *Semine minutissimo; flore nullo, vel imperfecto.* quae –
 1. *Cauliferae.* Lycopodium; Musci varij; Adiantha; Equiseta; Lichenes; Lunaria minor; Ophioglossum.
 2. "Ακαυλοι[99] Polypodium; Phyllitis; Filix; Lonchitis; Asplenium; Adiantha quaedam.

III. *Flore imperfecto, seu stamineo.* Lupulus; Cannabis; Urtica; Lapathum; Persicaria; Bistorta; Potamogiton; [...] Atriplex; Polygonum; Parietaria; Alchimilla; Blitum.

IV. *Flore composito, natura pleno lactescentes.* Lactuca; Sonchus; Hieracium; Dens Leonis; Tragopogon; Pulmonaria; Lampsana; Pilosella repens; Cichoreum.

V. *Flore composito, semine papposo non Lactescentes, flore discoide.* Tussilago; Petasites; Conyza; Jacobaea; Virga Aurea; Senecio; Gnaphalium.

VI. *Flore ex flosculis fistularibus composito*; sive Plantae **capitatae**. Carduus; Bardana; Cyanus; Jacea.

---

98. These systematic tables appear to have been compiled by Nicolson from Ray's *Historia Plantarum*, (1686–1688). Ray, in fact, published two works specifically on method: *Methodus Plantarum nova*, (1682), and *Methodus Plantarum emendata et aucta*, (1703). An account and assessment of these will be found in Raven, *Ray,* pages 192–200 and 287–94, but Nicolson himself uses an alphabetic arrangement for his *Catalogue* and gives no other indication that he was interested in systematics.

99. Lacking stalks.

VII. *Flore composito discoide, seminibus pappo destitutis*; **Corymbiferae** dictae. Ptarmica; Millefolium; Chamaemelum; Eupatorium Cannabinum; Scabiosa; Bellis; Tanecetum; Absinthium; Artemisia; Dipsacus; Eryngium.

VIII. *Flore perfecto, seminibus nudis solitarijs.* Valeriana; Linaria; Agrimonia; Pimpinella; Thalictrum; Fumaria.

IX. *Umbelliferae.* Quae vel –
1. *Semine lato,* seu *foliaceo.* Sphondylium; Pastinaca; Peucedanum.
2. *Semine longiore.* Pecten Veneris; Meum; Oenanthe Aquatica; Sium; Saxifraga; Foeniculum; Angelica Pimpinella saxi*fraga* Carum; Apium; Caucalis; Crithmum.
3. *Folijs integris.* Perfoliata; Auricula leporis; Sanicula.

X. *Stellatae.* Rubia; Cruciata; Gallium; Aparine.

XI. *Asperifoliae.* Cynoglossum; Echium; Buglossum; Lithospermum; Myosotis Scorpioides; Consolida.

XII. *Suffrutices,* et *Verticillatae.* Serpillum; Mentha; Pulegium; Marrubium; Verbena; Origanum; Clinopodium; Lamium; Sideritis; Bugula; Scordium; Hedera terrestris; Cardiaca; [...] Mentha; Betonica; [Lamium] etc.

XIII. *Semine nudo, polyspermae.* Ranunculus; Malva; Caryophyllata; Pentaphyllum; Tormentilla; Ulmaria; Anemone; Plantago Aquatica; etc.

XIV. *Bacciferae.* Chamaemorus; Bryonia; Ruscus; Polygonatum; Lilium convallium; Solanum; Vaccinia; Chamaerubus.

XV. *Multisiliquae,* seu *Corniculatae.* Sedum; Telephium; Rhodia Radix; Helleborus; Caltha palustris; etc.

XVI. *Fructu sicco singulari, flore monopetalo.* Hyoscyamus; Gentiana; Convolvulus; Campanula; Pinguicula; Linaria; Scrophularia; Digitalis; Euphrasia; Melampyrum.

XVII. *Flore tetrapetalo uniformi siliquosae.* Leucoium; Paronychia; Alliaria; Turritis; Rapistrum; Eruca; Cardamine; Cochlearia; Glastum; etc.

XVIII. *Vasculiferae, flore tetrapetalo, Anomalae et sui generis.* Veronica; Papaver; Lysimachia; Tithymalus; Plantago.

XIX. *Flore papilionaceo; seu Leguminosae.* Haeque vel –
1. *Scandentes.* Pisum; Lathyrus; vicia; Lens.
2. *Non trifoliatae, claviculis carentes.* Faba; Glycyrrhisa; Orobus; Astragalus; Anthyllis; Ferrum Equinum; Glaux.
3. *Trifoliatae.* Trifolia varia; Anonis; Lotus; etc.

XX. *Vasculiferae pentapetalae.* Et –
1. *Folijs in caule ex adverso binis.* Caryophyllus; Lychnis; Chamaecistus; Hypericum; Lysimachia lutea; Anagallis; Alsine; etc.
2. *Folijs in caule alterno, aut nullo ordine positis.* Sedum Alpinum; Cotyledon hirsuta; Gramen Parnassi; Viola; Geranium; Linum; etc.
3. *Flore pentapetaloide.* Primula veris; Centaurium; Verbascum; Acetosella; Nummularia; Ros solis; Valeriana Graeca; etc.

XXI. *Radice bulbosa.* Allium; Narcissus; Ornithogalum; Hyacinthus; Colchicum; Crocus.

XXII. *Bulbosis affines.* Iris; Arum; Asphodelus; Orchis; Helleborine; Bifolium; Pyrola.

XXIII. *Graminifoliae, flore imperfecto, Culmiferae.* Et –
1. *Grano majore;* quae *Cerealia dicitur*. Triticum; Secale; Hordeum; Avena.
2. *Grano minore; quae Gramina vocantur.* Et sunt vel –
   1. *Spicata.*
   2. *Paniculata.*
   3. *Avenacea.*

XXIV. *Graminifoliae non Culmiferae.* Gramen Cyperoides; Gr*amen* junceum; Sparganium; Acorus; Typha.

XXV. *Anomalae, et incertae sedis.*
1. *Aquaticae.* Nymphaea; Millefolium Aquati*cum*; Cotyledon aquatica; Stellaria; gladiolus Lacustris.
2. *Terrestres.* Persicaria Siliquosa; Myosuros; Polygala; Cuscuta.

XXVI. *Arbores* et *Frutices.* vel –
1. *Flore a fructu remoto.* haeque –
   1. *Nuciferae.* Juglans; Corylus; Fagus; Castanea; Quercus.
   2. *Coniferae.* Abies; Pinus; Alnus; Betula.
   3. *Bacciferae.* Juniperus; Taxus.
   4. *Lanigerae.* Populus; Salix.
   5. *Vasculis foliaceis.* Betulus.
2. *Fructu flori contiguo.* Et –

1. *Summo fructui insidente flore.* Malus; Pyrus; Sorbus; Mespilus; Rosa; Ribes; Vitis Idaea; Vaccinia rubra; Periclymenum; Hedera; Oxyacantha.
2. *flore imo fructui, cohaerente.* Et –
   1. *Fructu per maturitatem humido.*
      1. *Pruniferae.* Prunus; Cerasus.
      2. *Bacciferae.* Arbutus; Viscus; Sambucus [aquatica]; Laureola; Viburnum; Rhamnus *secundus* Clusij; Ligustrum; Alnus nigra; Berberis; Agrifolium; Rhamnus Catharcticus; Erica baccifera; Rubus; Euonymus Theophrasti.
   2. *Fructu per maturitatem sicco.* Staphylodendron; Buxus; Ulmus; Fraxinus; Acer; Gale; Erica; Genistae species o*mn*es; Tylia.

page 1 **A**

Abies. **Fir-tree.** 'ἐλάτη.[1] gyr-treow.
1. Alba, vel foemina. forte tamen Picea. Our moss-wood[2] in BRIGSTEER, WRAGMIRE, etc. — Brigsteer SD 482896; Wragmire NY 453493
2. Mas, Conis sursum spectantibus. Pl. Staff.

Abrotanum Campestre vel Sylvestre. **Wild Sothern-wood.** Isle of WALNEY. — *Artemisia campestris* L.

Absinthium. **Wormwood.**
1. Vulgare, Latifolium. **S** — *Artemisia absinthium* L.
2. Seriphium vel Marinum Near CARTMEL-Grainge.[3] Hindpull, nigh the Isle of WALNEY. — *Artemisia maritima* L.; Hindpool SD 193696

Acanthium vulgare, Album. **Common Cotton-Thistle.** 'Twixt NEWCASTLE and the Glass-Houses.[4] SOUTH-SHEELDS in Com*itatu* Dun*elmense*. — *Onopordum acanthium* L.

page 2 Acer. **The Maple-Tree.**
1. Majus, Latifolium. Quibusdam (falso) Sycomorus et Platanus.
2. Minus. At CROSS-BANK[5] near KENDALL; By the Elm-Tree. MELKIN-THORP. about RIPPON etc. abundantly. — *Acer campestre* L.; Melkinthorpe NY 556252

---

1. pine tree.
2. Like Gerard before him (*The Herball,* (1597), page 1181), Nicolson thought that the remnants of the ancient British pine forests (*Pinus sylvestris*) which are sometimes dug out of bogs and mosses were *Picea abies* or possibly *Abies alba.* Unlike Gerard, however, Nicolson was reluctant to believe that "all our Moss-wood" had been overthrown at the time of the Deluge (Letter to John Morton, 17th June, 1706: *Bodleian Lib MS Addl.* C 217, fol. 37) as he found it difficult to believe that the wood could have been preserved for so long.
3. The Prestons, who were lords of Holker, acquired much of the land that had belonged to Cartmel Priory, including Grange, the monastic farm that was to become the Victorian holiday resort of Grange-over-Sands. See Sam. Taylor, *Cartmel People and Priory,* (Kendal, 1955), page 36.
4. The name "Glasshouse Bridge" over the Ouse Burn (NZ 264642) is now the only reminder of the glassmaking industry, which was flourishing in the mid-17th century as a monopoly of Sir Robert Mansell (1573–1656). The area is now much built over.
5. Crossbank is marked on John Todd's map of Kendal, 1787, as being a hamlet between the Penrith road and the Appleby road in a position SD 522935.

Acetosa. **Sorrel.**

1. Vulgaris vel pratensis; folio longo. **S** AS. Gaeces sure i.e. Cuckow-Sorrel.	*Rumex acetosa* L.
2. Minor, Auriculata repens. **Sheeps Sorrel. S**	*Rumex acetosella* L.
3. Rotundifolia repens Westmorlandica. T. L. Hortensis. C. B. Syn p. 316. BUCK-BARROW-WELL.	*Oxyria digyna* (L.) Hill Buckbarrow Well NY 478076
4. Acetosella. Vide Trifolium.	

Acinos Anglica. **Small Stone-Basil.**
Vide Clinopodium.

Aconitum racemosum.
Vide Christophoriana.

Acorus. **Flag**, or **Segg.**

page 3

1. Verus; sive Calamus (falso) Aromaticus.	
2. Iris Palustris Lutea. **Water-Flowerdeluce. S**	*Iris pseudacorus* L.

Adiantum. **Maiden-hair.** Vide Dryopteris.

1. 'Ακρόστιχον[6] seu furcatum. In EDINBURG-Park. J.R.	*Asplenium septentrionale* (L.) Hoffm.
2. Album; vel Ruta muraria. **S**	*Asplenium ruta-muraria* L.
3. folio filicis. i.e. Dryopteris alba.	
3. Album Crispum Alpinum. Scotiae, J.R.	*Cryptogramma crispa* (L.) R.Br. ex Hook
4. Album floridum. J.R. ORTON in WEST-MERLAND.	*Cryptogramma crispa* (L.) R.Br. ex Hook Orton NY 623083
4. Majus Coriandri folio.	
5. Aureum. Cujus *sunt* Species –	
1. Majus. **Great Goldilocks.** In Bogs at WASDALE-head[7]	*Polytrichum commune* L.
2. Minus. In Muris et Aggeribus **S**	*Polytrichum juniperum* Willd.

6. divided at the end, forked, as glossed by *furcatum*. This was a record of the itinerant plant collector, Thomas Willisel, who noted it from Salisbury Crags in Holyrood Park where it still grows (Harold R. Fletcher and William H. Brown, *The Royal Botanic Garden Edinburgh 1670–1970*, (Edinburgh, 1970), page 10).
7. Not, however, the Wasdale Head of Central Lakeland, but the farm of that name, now abandoned, just off the present A6 south of Shap at NY 550082.

	3. Humilius, folijs latis subrotundis. **S**	prob. *Polytrichum formosum* Hedw.
	4. Petraeum perpusillum, folijs bifidis vel trifidis. T.L. Wrenose.	*Hymenophyllum wilsonii* Hook Wrynose Pass NY 278027
page 4	AEgilops bromoides. Vide Gramen festucae, seu; Avena pilosa.	
	Agrifolium, vel Aquifolium. **Holly-Tree.**	
	1. Vulgaris. **S**	*Ilex aquifolium* L.
	2. Baccis luteis, nondum descriptum.	
	3. Folio variegato. Specie (forte) non differunt, sed Accidentaliter. J.R. Omisit ergo in *Synopsi Meth.*	
	Agrimonia vulgaris. **Agrimony. S**	*Agrimonia eupatoria* L.
	Alcea vulgaris, vel Malva verbenacea. **Vervain-Mallow. S** flore rub*ro*.	*Malva moschata* L.
	Alchimilla. **Ladies Mantle.**	
	1. Vulgaris; Pes Leonis. J.B. Nostratibus, **Bearsfoot. S**	*Alchemilla vulgaris* agg.
	2. Alpina pentaphyllos. J.R. et T.L. Buckbarrow-well. etc.	*Alchemilla alpina* L. Buckbarrow Well NY 478076
	3. Montana minima. i.e. Percepier Anglorum.	
page 5	Alectorolophos. **Vide Pedicularis.**	
	Alga. **Wrack.** Vide Fucus.	
	1. Marina platyceros porosa. **Silken Wrack.**	
	2. Trichodes fontalis. **Water-Maidenhair.**	
	Alliaria. **Jack by the Hedge,** or **Sawce alone. S**	*Alliaria petiolata* (Bieb.) Cavara & Grande
	Allium. **Garlick.**	
	7. Amphicarpon. Th. Lawson.	
	1. Montanum bicorne. BRADLEY-FEILD; nigh KENDALL. Rocks about SETTLE.	prob. *Allium oleraceum* L. Bradleyfield SD 493917

2. Sylvestre bicorne. Qu. Annon idem cum Superiore? J.R.	
3. Ursinum. **Ramsons**, or **Ramps. S**	*Allium ursinum* L.
4. Sylvestre. Ger. emac. **Crow-Garlick. S**	*Allium vineale* L.
5. Holmense Sphaericeo Capite.	
6. Mont*anum* purp*ureum* proliferum. MABURG. SETTLE, etc.	*Allium carinatum* L. Mayburgh[8] NY 519284

page 6

Alnus. **Alder-Tree.** AS. Aelre.	
1. Vulgaris; Rotundifolia glutinosa viridis. **S**	*Alnus glutinosa* (L.) Gaertn.
2. Nigra baccifera. [T.L.] WHINFIELD.[9] CUNSWICK-SCAR. In most of the little Holms[10] in WINDERMERE.	*Frangula alnus* Mill. Whinfell Forest NY 573275 Cunswick Scar SD 491940

Alopecuros maxima Anglica. Great **Marsh-Foxtail-Grass.** By BROUGHAM-Church.	*Polypogon monspeliensis* (L.) Desf. Brougham Church NY 527284

Alsine. **Chickweed** Vide Spergula. Saxifraga.	
1. Serrato folio glabro. **Germander-Chickweed.** in Hortis.	*Veronica agrestis* L.
2. Folijs subrotundis Veronicae. **S**	*Veronica arvensis* L.
3. Hederacea. **Ivy-chickweed** or **Henbit.** KENDAL-CASTLE. Below WATER-FALL-BRIDGE.	*Veronica hederifolia* L. Kendal Castle SD 522924 Waterfalls Bridge NY 551240
4. Hirsuta Myosotis. **Mouse-ear-Chickweed. S**	*Cerastium glomeratum* Thuill. or *C. holosteoides* Fr.

---

8. For the etymology of the name, see pages 210–12 of C. W. Dymond, "Mayburgh and King Arthur's Round Table", *CW1* xi 187–219. Also *PNW* ii 205.
9. *PNW* ii 132.
10. i.e. islands. The word is still preserved in the Windermere islands Lady Holme, Ramp Holme and others. It is also used of land or fields temporarily enisled by floodwater. See below n 11 page 10.

page 7	[5. Myosotis facie Lychnis Alpina flore amplo niveo Vide J.R. fasc Stirp.]	
	6. Hirsuta altera Viscosa.	
	7. Longifolia, uliginosis proveniens locis. **S**	*Stellaria alsine* Grimm.
	8. Minor Multicaulis. **S**	*Arenaria serpyllifolia* L.
	9. Palustris; folijs tenuissimis. Vide Saxifraga pal*ustris*.	
	10. Palustris major perennis.	
	10. Plantaginis folio.	
	11. Rotundifolia; vel Portulaca Aquatica. **S**	*Peplis portula* L.
	12. Spuria, pusilla, repens; folijs Saxifragae aureae.	
	13. Pusilla, pulchro flore, folio tenuissimo nostras. J.R. fasc.	
	14. Vulgaris, sive Morsus Gallinae. **S**	*Stellaria media* (L.) Vill.
	Althaea vulgaris. **Marsh-Mallowes.**	
page 8	Alysson Germanicum Echioides. i.e. Buglossum sylvestre caulis prominentibus. **Great Goose-grass. S**	*Asperugo procumbens* L.
	Anagallis. **Pimpernel.**	
	1. Mas; flore phoeniceo. **S** Among the corn in the Holm,[11] etc.	*Anagallis arvensis* L.
	2. Foemina; coeruleo flore.	
	3. Lutea nemorum. maxime vulgaris, in Bogs, Common.	*Lysimachia nemorum* L.
	4. Aquatica. **Brooklime.** quae vel –	
	1. Vulgaris. Becabunga officinarum. **S**	*Veronica beccabunga* L.
	2. Minor, folio oblongo, flore purpurascente.	
	3. Rotundifolia. 'Twixt WITHERSLACK and GRAINGE, in ditches. flore albo.	*Samolus valerandi* L. Witherslack SD 435848 Grange-over-Sands

11. A "holm" may be either an island (see above n 10 page 9) or a "piece of flat low-lying ground by a river or stream, submerged or surrounded in time of flood" (O.E.D.). The holm lands of Great Salkeld lay along the bank of the Eden and are among the lands mentioned as belonging to Ranulph Meschin in the Middle Ages (A. G. Loftie, *Great Salkeld, its Rectors and History*, (1900), page 25). The Tithe map of Great Salkeld Parish, 1840, shows the name preserved in fields such as Low Holm, North Holm and others.

4. Angustifolia. **S** Nigh Long-*Sleddal-Chappel.* flore albo.	*Veronica scutellata* L. Longsleddale Chapel NY 501029
Anchusa degener facie Milij Solis. **Bastard Alkanet, Gromil** or **Salfern.** Lancemoor near NEWBY in Westm*orland.*	*Lithospermum arvense* L. Lansmere NY 575217
Androsaemum. **Tutsan** or **Park leaves.**	
1. Vulgare; maximum frutescens. Bellingham's-Holm[12] in WINANDERMEER. RYDALL. Cart-mel-well.[13] Marsh-Grainge.	*Hypericum androsaemum* L. Marsh Grange[14] SD 220797
2. Magnum; sive Ruta Sylv*estris* Hypericoides. **Great St John's-wort.** Johnson	
Anemone. **Anemony** or **Crow-foot.**	
1. Nemorum, flore majore albo. **S**	*Anemone nemorosa* L.
2. Tuberosa radice. **Windflower.**	
Angelica.	
1. Sylvestris major. Idem quod Archangelica. J.R. In the Hedges about CARLILE. **S**	*Angelica sylvestris* L.
2. Minor, sive Erratica. **Herb Gerard, Goutwort** or **Ashweed.**	

page 9

12. This is presumably the island granted to Alan Bellingham Esquire by Henry VIII and formerly called Roger Holme (*Nicolson and Burn*, vol i, page 186). The grid reference is SD 399979 and the modern name is Rough Holme (See W. G. Collingwood, *The Lake Counties*, new edition, (London and NY, 1932), page 353).
13. This, sometimes referred to as "Cartmel medicinal well", is Holy Well Spa on Humphrey Head (SD 390738). Spas were becoming fashionable in the late 17th century: Nicolson's contemporary, Celia Fiennes, visited many in her travels about England (ed C. Morris, *The Journeys of Celia Fiennes*, revised edition, (1949), page xxvi) and the Holy Well Spa was described in Charles Leigh, *The Natural History of Lancashire, Cheshire, and the Peak, in Derbyshire*, (Oxford, 1700), page 50.
14. In the 16th and 17th centuries, Marsh Grange was the home of the Askew family. Margaret Askew, the most famous member of the family, married Judge Thomas Fell of Swarthmoor Hall and became a central figure in the Quaker movement after her conversion by Fox in 1652. In the late 17th century, Marsh Grange became successively the home of Margaret Fell's son George and her daughter Mary, married to Thomas Lower (Isabel Ross, *Margaret Fell*, (London, NY and Toronto, 1949), page 5). Thomas Lawson, whose association with the movement also dated from Fox's visit, gave botanical instruction to several members of the family (Maria Webb, *The Fells of Swarthmoor Hall and their Friends*, (1865), pages 371–72).

page 10

Anonis. **Rest-Harrow, Cammock** or **Petty-whin.** Κολοβοανθος'.[15] Theophr.

1. Spinosa, flore purpureo. At Hall-head, at the head of CUNSWICK-Scar. — *Ononis spinosa* L. Halhead Nab SD 492944
2. Spinis carens, purpurea. **S** — *Ononis repens* L.

Anthyllis.

1. Leguminosa praten⟨s⟩is Vulneraria. **Kidney-vetch, Ladie's finger. S** — *Anthyllis vulneraria* L.
2. Maritima Lentifolia. **Sea-pimpernel.**

Antirrhinum. **Snapdragon** or **Calves-snout.**

1. Majus.
2. Mi⟨nim⟩um, Repens.

page 11

Aparine. **Goose-grass.**

1. Major. Vide Alysson Germ*anicum.*
2. Vulgaris. Lappago Plinij. **S Cleavers.** — *Galium aparine* L.
3. Semine laeviore.

Aphaca.\folio deltaoide./ **Yellow Vetchling. S** At [ye Willies bridge . . .[16]]
In the foot-way 'twixt KENDALL and Hall-guards. — *Lathyrus aphaca* L. Hallgarth SD 507938

Apium. **Smallage**. In paludosis, praesertim maritimis. About CARTMEL-well.[17] — *Apium graveolens* L.
marinum, Scoticum. near EDINBURG. J.R. ex SIBBALD.[18] — *Ligusticum scoticum* L.

Aquilegia sylvestris. **Columbines. S** — *Aquilegia vulgaris* L.

Aracus. Vide Vicia.

---

15. Nicolson, in humanistic fashion, has made a noun from the adjective κολοβοανθης, bearing stunted flowers.
16. See below n 44 page 23.
17. See above n 13 page 11.
18. Robert Sibbald (1641–1722), together with Andrew Balfour (1630–1694), was responsible for the establishment of the botanic garden at Edinburgh, now the Royal Botanic Garden. Nicolson speaks highly of the two men as the founders of Natural History in Scotland and as "two of the greatest Ornaments of their Country and Profession which this Age has produc'd" (*The Scottish Historical Library*, (1702), page 26). *Ligusticum scoticum* was, in fact, first recorded from Britain in Sibbald's *Scotia Illustrata*, (1684), (See Fletcher and Brown, *The Royal Botanic Garden*, page 10).

Arbutus. **Straw-berry-Tree**. In the west of Ireland. J.R.	*Arbutus unedo* L.	
Arctium montanum. **Button-Burr.**		
Argemone. **Popp[e]y.**		page 12
1. capitulo longiore spinoso.		
2. Cambro-Britannica Lutea, perennis. Milthrop.	*Meconopsis cambrica* (L.) Vig.[19] Milnthorpe	
3. Capitulo breviore hispido. **S**	*Papaver hybridum* L.	
4. Capitulo breviore glabro.		
Argentina. **Wild Tansie, Silver-weed. S**	*Potentilla anserina* L.	
Aria Theophrasti. **White beam-tree**. [T.L.] Vide Sorbus. In LEVENS-park.	*Sorbus aria* agg. Levens Park SD 502855	
Armeria.		
1. Pratensis, flore\interdum/ Albo. T.L. **Meadow-pink, Wild Williams, Cuckow-flower, Crow-flower. S**	*Lychnis flos-cuculi* L.	
2. Caliculo foliolis fastigiatis cincto. **Deptford-Pink** of*ficinarum*.		
Artemisia Vulgaris. **Mugwort. S**	*Artemisia vulgaris* L.	page 13
Arum vulgare. **Cuckow-pint**, or **Wake-Robin**.		
1. maculatum, maculis candidis vel nigris. **S**	*Arum maculatum* L.	
2. non maculatum. **S**	*Arum maculatum* L.	
Arundo vallatoria, vulgaris palustris, **Common Reed.** *Κάλαμος χαρακίας*.[20] Theoph.		
Ascyrum. **St Peter's wort.**		
1. Vulgare, Caule quadrangulo. **S**	*Hypericum tetrapterum* Fr.	

19. J. A. Martindale in "Early Westmorland Plant Records", *The Westmorland Natural History Record*, (London and Kendal, 1888–89), vol i, pages 67–68, raises the question of whether, in the 17th century, this was a Westmorland plant in view of the fact that neither Lawson nor Ray record it from the county, although Hudson, writing in 1762, notes it as plentiful around Kendal. This record establishes it as having reached South Westmorland by the end of the 17th century.
20. Reed fit to be used as a stake.

2. Supinum villosum palustre. Rotundifolium. 'Ελώδης.[21] In Boggs near WINANDER-MEER. **S** — *Hypericum elodes* L.

page 14 Asparagus. **Asparagus** or **Sperage.**

1. Vulgaris. Johnston.
2. Palustris, vel marinus; folio crassiore. On IRELITH-MARSH[22] in FOURNESS. In ULVERSTON-moss.[23] — *Asparagus officinalis* L. ssp. prostratus (Dum.) E.F. Warb.

Asperula. **Wood-roof.**

1. montana odorata.
2. Coerulea Arvensis. **S** — *Sherardia arvensis* L.

Asphodelus. **Asphodil.**

1. Bulbosus. Vide Ornithogalum Angustifolium.
2. Palustris Scoticus, Minimus. Near BERWICK.[24] — *Tofieldia pusilla* (Michx.) Pers.
3. Lancastriae. In CLIBBURN-MOSS, RENWICK-HOLMS,[25] etc. — *Narthecium ossifragum* (L.) Huds. Cliburn Moss NY 576257

page 15 Asplenium vel Ceterach. Scolopendria vera. **Ceterach, Spleenwort, Miltwast.** On MAIDEN-TOWER 'twixt KENDALL and Brigsteer. TROUTBECK-bridge. — *Ceterach officinarum* DC. Maiden-Tower not located. Troutbeck Bridge NY 404003

Aster. Vide Tripolium.

---

21. Marshy.
22. Ireleth Marsh, west of Ireleth, is shown on the O.S. one-inch map of 1864 over an area now occupied by the village of Askam in Furness (SD 213775).
23. Marked on the O.S. six-inch map of 1850 at SD 302785.
24. This is a Raian location. George Johnston, *A Flora of Berwick-upon-Tweed*, (Edinburgh and London, 1829–31), vol i, page 83 reported searching in vain for this plant and wondered whether, perhaps, North Berwick were intended. In Ray's northern tour of 1671, however, he reports visiting "Scrammerston Mill", which is about a mile and a half south of Berwick-on-Tweed on the day following his discovery of *Tofieldia pusilla* (ed. Edwin Lankester, *Memorials of John Ray*, (1846), page 151).
25. This is probably Mill holme, a meadow bordering the river on the Tithe map of Renwick Parish, circa 1844, at NY 595428. However, an island is shown in the river here and may be the holm referred to (see n 11 page 10).

Astragalus Sylvaticus. **Wood-pease** or **Heath-pease. S** Scotis, KAREMYLE.[26] Vide Sibbald. p. 43. — *Lathyrus montanus* Bernh.

Atriplex. **Orach**, or **Notchweed**.
1. Canina vel olida; juxta semitas. **Stinking orach.** L.[27] — *Chenopodium vulvaria* L.
2. Sylvestris. [Angustifolia].
   1. Angustifolia. BIGGER in the Isle of WALNEY. — *Atriplex patula* L. Biggar SD 191662
   2. Vulgaris; folio Sinuato. **S** — *Chenopodium album* L.
   3. Sylvestris Altera.
   4. Latifolia. **Goose-foot**, or **Sowbane.**
3. Maritima, quae
   1. Angustifolia.
   3. Fruticosa Vide Halimus.
   2. Repens, Laniciata.

Avena. **Oats**. AS. Atan.

page 16

1. Vulgaris, vel Alba. **S** — *Avena sativa* L.
2. Nigra. Rarior. **S** — *Avena sativa* L.
3. Nuda, Cornubiae. **Pill-Corn**.
4. Pilosa. Haver.

Auricula Leporis, folio angustissimo. **The least Hares-Ear**.

Auricula muris. **Mouse-Ear**. Vide Alsine Hirsuta.
1. flore Albo majore. About[28] SCARBROUGH-castle. On the Skirts of the WOLDS. — *Cerastium arvense* L. Scarborough Castle TA 049893

---

26. Geoffrey Grigson, *The Englishman's Flora*, (1958), page 142 gives "Cormeille" and "Corra-meile" as names from the north of Scotland and the Hebrides, where the tuberous rhizome was dug up and eaten as described by Sibbald who equates it with the food of the Ancient Britons mentioned by Dio Cassius in his life of Severus the emperor.
27. I think it extremely unlikely that the letter "L" here refers to a person – it is not an abbreviation used by Ray, and Nicolson's practice was to give names or full initials. I suggest, rather, that it may have stood for "Lowther". We know that Nicolson visited these gardens and all the four plants against which he puts "L" are unlikely to have been found outside of cultivation. Traveller's Joy (page 93) and Gold of Pleasure (page 62) were noticed by Nicolson on a visit in 1690 (see pages xxxvi and xxxvii) and Stinking Orach grew not far away at Hutton-in-the-Forest. Hornbeam was planted in Lowther woods.
28. The letter "A" in "About" has been written over an S for Salkeld.

2. Pulchro flore, folio tenuissimo. Among the Tenters[29] at KENDALL. On the mountains about SETTLE.
Minuartia verna (L.) Hiern

B

page 17 Baccharis Monspeliensium. Vide Conysa major. T.L.

Ballote. i.e. Marrubium nigrum, foetidum. **S** About the walls at CARLILE.
Ballota nigra L.

Barbarea. **Winter-Cresses**, or **Rocket**.
1. Vulgaris, Lutea latifolia. **S**
Barbarea vulgaris R. Br.
2. Muralis. **Wall-Cress, with Daisy-Leaves. S**
Arabis hirsuta (L.) Scop.

Bardana. **Bur-dock.**
1. Major: sive Ar\c/tium Dioscoridis.
1. Major \altera/ Lanuginosis Capitulis. **S**
Arctium lappa L.
2. Minor. Rarior. J.R. Vide Arctium montanum.

page 18 Becabunga. i.e. Anagallis Aquatica.

Behen Album, officinarum. **Spatling Poppey**, or **White bottle. S** Bladder-Campion.
Silene vulgaris (Moench) Garcke
Rubrum. Vide Limonium.

Bella Donna. Vide Solanum Lethale.

Bellis. **Daisie**, or **Ox-eye**.
1. Major Sylvestris. **S**
Chrysanthemum leucanthemum L.
2. Minor. Ubique in pratis et pascuis. **S**
Bellis perennis L.

Berberis officinarum. Vide Oxyacantha.

---

29. Kendal's woollen industry, established in the 14th century, had reached its highest repute in the early 17th century and areas around the town were used for the "tentering" or hanging and stretching of cloth. John Todd's map of 1787 marks "Tenter Fell" north-east of the Ambleside road opposite the House of Correction (SD 513935) but tenter hooks are also depicted on the southern side of Kendal Fell (SD 507926), an area still known locally as Tenterfell.

Beta Sylvestris, Maritima. **Sea-Beet.**

Betonica. **Betony**. AS. Attor-lathe. — page 19

1. Vulgaris, flore purpureo. **S** — *Betonica officinalis* L.
2. flore albo. T.L.
3. Aquatica. Vel Ocymastrum majus. Nether- and upper- LEVENS. At the Gates at RYDALL.[30] — *Scrophularia aquatica* L. Nether Levens SD 488851 Levens SD 488862
4. Pauli. i.e. Veronica pratensis minor.

Betula. **Birch-tree. S** — *Betula pubescens* Ehrh. and *B. pendula* Roth

Betulus. **Horn-beam-tree, Horse-beech**, or **Horn-beech**. L.[31] — *Carpinus betulus* L.

Bifolium. **Tway-blade**.

1. Sylvestre
   1. Majus. **S** — *Listera ovata* (L.) R.Br.
   2. Minimum. On several Heaths in YORKSHIRE, LANCASHIRE and NORTHUMBERLAND. J.R. **S** — *Listera cordata* (L.) R.Br.
2. Palustre.

Bistorta. **Bistort**, or **Snake-weed.** — page 20

1. Vulgaris major.[(a)] — *Polygonum bistorta* L.[32]
3. Minima Alpina. — *Polygonum viviparum* L.
2. Minor Nostras.\J.R. et-/T.L.\(a)/ Septent*rionalis*. **Eastermintgiants**; corrupte pro **Eastern-Magicians**. At — *Polygonum bistorta* L.

30. It was usual, at a time when the land outside villages was largely unenclosed, to have gates across the roads entering. An estate map of Rydal of 1770 (WD/Ry, Records Office, Kendal) shows several such gates and Loftie, in his history of Great Salkeld, page 109, writes interestingly of the gates that protected Nicolson's own village from the attentions of wandering livestock.
31. See above n 27 page 15.
32. Nicolson would appear by his notation (the two superscribed 'a's) to equate "Bistorta vulgaris major" with "Bistorta minor nostras". Both are undoubtedly *Polygonum bistorta* for which Eastermintgiants is a Northern name deriving, according to Geoffrey Grigson, *The Englishman's Flora*, (1958), page 230 from Easter + manger = the food eaten at Easter. There seems to have been some overlap of the *Polygonum* spp in Nicolson's mind as there was in Ray's – Ray's "Bistorta minor nostras" (*Synopsis 1724*, page 147) which he records from Crosby Ravensworth where it had been known since the time of Gerard, 1597 (*The Herball*, page 323), is more probably *Polygonum viviparum*. The last bistort in Nicolson's list could, I think, be either *P. bistorta* or *P. viviparum*.

KIRK-OSWALD CASTLE; LANGWATHBY-Holm,[33] etc. — Kirkoswald Castle NY 560410
Minor. On TEES. Near DUMMA-HILL juxta STAINMORE.[34] — Dummah Hill NY 827154

Blattaria major flore luteo. **Great Moth Mullein**. Johns.

Blitum Album minus. **Upright Blite**, or **All-Seed**. Vide Atriplex. Kali. Bonus Henricus. Vermicularis.

Bonus Henricus vel Tota bona. **Common Mercury**, or **Al-good. S** — *Chenopodium bonus-henricus* L.

Branca Ursina. Vide Sphondylium vulgare.

Brassica. **Colewort.**
1. Marina monospermos Anglica. Common on our Shore. — *Crambe maritima* L.
2. Sylvestris.
3. Arborea Morisoni.

page 21 Bromos sterilis. Vide Festuca.

Bryon Lactucae folijs. Vide Lichen Marinus.

Bryonia. **Bryony.**
1. Vulgaris Alba; quae et vitis alba d*icitu*r. AS. hwit wilde wingeard Near LANCASTER. ULNABY in Com*itatu* DUNELM*ense*. — *Bryonia dioica* Jacq. Ulnaby Hall NZ 227172
2. Nigra. Sigillum B*eatae* Mariae officinarum. **Lady's Seal**. AS. blac wingeard. About KENDAL, copiose. — *Tamus communis* L.

Buglossum. **Bugloss.**
1. Sylvestre minus. **S** — *Anchusa arvensis* (L.) Bieb.

---

33. On the Tithe map of Langwathby Parish, 1839, the holm lands are shown as lying in a bend of the R. Eden, west of the bridge at NY 562338.
34. *Minor ... Stainmore.* This was a late addition to the MS, squeezed in as a side note.

2. Luteum. Lang de boeuf. 'Twixt STOCKDON and NORTON in Com*itatu* DUNELM*ense*. — *Picris echioides* L. Stockton Norton NZ 446218

Bugula. **Bugle.**
1. Vulgaris Pratensis. **S** — *Ajuga reptans* L.
2. Coerulea Alpina.

Bulbocastanum. **Earth-Nut**, **Ear-Nut** or **Kippernut. S** — *Conopodium majus* (Gouan) Loret

page 22

Bulbocodium. Vide Narcissus.

Bunias Sylvestris. Vide Napus.

Bupleurum. Vide Auricula Leporis.

Buphthalmum vulgare. Quibusdam, Chamaemelum Chrysanthemum. **Common Ox-Eye.** [**S**] Near SOGBURN in Com*itatu* Dunelm*ense* J.R. — *Anthemis tinctoria* L. Sockburn NZ 347073

Bursa pastoris. **Shepherd's purse.**
1. Major vulgaris. **S** — *Capsella bursa-pastoris* (L.) Medic.
2. Majori Affinis, loculo sublongo. In CRAVEN. J.R. — *Draba muralis* L.
3. Minor, folijs sinuatis. **S** — *Teesdalia nudicaulis* (L.) R.Br.

Buxus. **Box-tree**.

C

Calamintha. **Calamint**.

page 23

1. Vulgaris officinarum.
2. Odore Pulegij, flore minore.[35] At KENDALL-CASTLE. — Kendal Castle SD 522924
3. Aquatica. **S** common. — *Mentha arvensis* L.

---

35. John Wilson, *A Synopsis of British Plants, in Mr Ray's Method*, (Newcastle-on-Tyne, 1744), page 97, independently identified the plant at Kendal Castle as *Calamintha nepeta* but Martindale, *The Westmorland Natural History Record*, page 117, disputes this find and it may be that a mis-identification for *Calamintha ascendens* was made by both men.

Page	Entry	Modern name
	Calceolus Mariae. **Lady's Slipper.** HELKS-WOOD[36] by INGLE-BORROUGH.	*Cypripedium calceolus* L.
	Calcitrapa. Vide Carduus Stellatus.	
	Caltha palustris. **Marsh-marigold. S**	*Caltha palustris* L.
	Camelina. **Treacle-wormseed**.	
	Campanula. Vide Trachelium.	
	1. Rotundifolia. **S**	*Campanula rotundifolia* L.
	2. folio Hederaceo.	
page 24	Cannabis. **Hemp**.	
	1. Sativa, vulgaris.	
	1. Mas. **S**	*Cannabis sativa* L.
	2. Foemina. **S**	*Cannabis sativa* L.
	2. Spuria. **Nettle-Hemp**.	
	1. flore albo. **S**	*Galeopsis tetrahit* agg.
	2. flore Luteo. [T.L]. ORTON and BURTON.	*Galeopsis speciosa* Mill. Orton NY 623083 Burton SD 530765
	Cannabina Aquatica. Vide Eupatorium Cannabinum.	
	Capillus Veneris. **Common Maidenhair**. Johns. **S**	*Asplenium trichomanes* L.
	Caput Gallinaceum Belgarum. Vide Onobrychis.	
page 25	Caprifolium. Vide Periclymenum.	
	Cardamine. **Ladies-Smock**, or **Cuckow-flower**.	
	1. Vulgaris, ubique in Pratis. **S**	*Cardamine pratensis* L.
	2. Impatiens. Duplex, J.R. **S** minima. Under the Scars at WHERF near SETTLE. J.R.	*Cardamine impatiens* L. Wharfe SD 784696
	3. Flore pleno.T.L.	*Cardamine pratensis* L.
	4. Pumila Bellidis folio, Alpina.	

36. Lady's Slipper Orchid, known from this location since the time of Parkinson, 1640, had disappeared by the late 18th century. Its disappearance was described by F. A. Lees from Withering's *Botanical Arrangement*: "I searched for it in vain in Helks Wood, a gardener of Ingleton having eradicated every plant for sale; *Mr. Woodward*" (Lees, *The Flora of West Yorkshire*, (1888), page 436).

Entry	Identification	Page
Cardiaca. **Motherwort**. Inter Rudera. As, in the Abbey at Carlile.[37]	*Leonurus cardiaca* L.	
Carduus. **A Thistle**. Vide Cirsium. Acanthium.		
1. Carlina acaulis minor. **Dwarf Carline-Thistle**. KENDALL-fell. SHAP.	*Cirsium acaulon* (L.) Scop.[38] Kendal Fell summit SD 504930	
2. Arvensis major, flore purpureo. **Musk Thistle**.		
3. Lacteus, \vel Mariae./ **Milk Thistle**, or **Ladies Thistle**. About APPLEBY.	*Silybum marianum* (L.) Gaertn.	page 26
4. 1. albis maculis notatus. 'Twixt CARLTON-town and the Hall.	*Silybum marianum* (L.) Gaertn. Carleton NY 530298 Carleton Hall NY 526293	
2. Non maculatus.		
4. Lanceatus. **Spear Thistle. S**	*Cirsium vulgare* (Savi) Ten.	
5. Nutans, tota stirpe Spinosus.		
6. Monstrosus Imperati. T.L.		
6. Moschatus. Vide Nutans.		
7. Polyacanthos. **Thistle upon Thistle.** [item cum . . .] TROWGILL.	*Carduus acanthoides* L. Trough Gill NY 587240	
8. Onopyxos vel Asininus **Asses Box**. Johns.		
9. Palustris. Ad hominis altitudinem.		
10. Stellatus. **Star-Thistle**.		
11. Tomentosus, vel Corona fratrum. **Woolly headed Thistle.**		
12. Vulgatissimus viarum. **S**	*Cirsium arvense* (L.) Scop.	page 27
Caryophyllata, vulgo Herba Benedicta. **Herb Bennet**.		
1. Vulgaris, ad Sepes. **S**	*Geum urbanum* L.	
2. Vulgaris major. About the HOLM,[39] RABY-COAT, etc.	*Geum rivale x urbanum* Raby Cote NY 180524	

37. The Cathedral at Carlisle is built on the site of the old priory which, from very early times was known as the Abbey, although the church never had an Abbot distinct from the Bishop (*V.C.H., Cumberland*, vol ii, 1905, page 135). There is still an Abbey Street, skirting the Cathedral precinct on the south-west side.
38. Probably a mis-identification for *Carlina vulgaris*. See also J. G. Baker, *A Flora of the English Lake District*, (1885), page 130–31.
39. See above n 11 page 10.

3. Montana purpurea. T.L. apud J.R. in Fasc. On the banks of the River KENT. SETTLE, INGLETON, etc.
   Geum rivale L.
   Ingleton SD 695732

Caryophyllus. i.e. Armeria. **A Pink**. Vide Holosteum.

1. Virgineus. **Maiden-Pink**. T.L. At Common-holm-bridge near Great STRICKLAND.
   Dianthus deltoides L.
   Commonholme Bridge NY 576247
2. Pumilio Alpinus.
3. Marinus minimus. **Thrift**, or **Sea-gilliflower**. Johns. On INGLEBOROUGH. and all along *ou*r Sea-coasts. above GARRY-GILL, 30 miles from the Sea.
   Armeria maritima (Mill.) Willd.
   Garrigill NY 745415

Carpinus. Vide Betulus.

Carthamus. Vide Cnicus.

Carum, vulgare. **Carawaies**. In Palustribus juxta HULL. J.R.
Carum carvi L.

page 28 Castanea vulgaris. **Chesnut-Tree**.

Catanance. **Crimson-grass-vetch**. 'Twixt the Glass-houses and Dent's-hole[40] (the foot-way) near NEWCASTLE.
Lathyrus nissolia L.

Caucalis. **Parsley**.

1. Anglica, flore rubente.
2. Minor, flosculis rubentibus. **Hedge-parsley. S**
   Torilis japonica (Houtt.) DC.
3. Segetum.
4. Nodosa, echinato semine. **Knotted parsley**.
5. Tenuifolia, flosculis subrubentibus.

page 29 Cauda Equina. Vide Equisetum.

Centaurium. **Centory**. *Κενταύρειον* et *Κενταύριον*.[41]

---

40. Dents hole is marked on G. Collins' chart of the River Tyne, 1753, on the north side of the river, opposite Friars Goose (NZ 277632). For a note on the glasshouses, see n 4 page 6.
41. These are simply alternative forms of the word for centaury.

1. Minus, vulgare. Above the Safe-guard at WETHERAL,[42] etc. **S** at CRAKE-LEEK-HOLM,[43] Willies-grassing,[44] etc.	*Centaurium erythraea* Rafn
2. Luteum perfoliatum.	
3. Palustre Luteum minimum.	

Cerasus. **Cherry-Tree**.

1. Sylvestris. Quae vel –	
1. Vulgaris, fructu rubro.	
2. fructu nigro. **Black Cherries**.	
3. Fructu minimo cordiformi. T.L. et Fasc. ROSGIL. etc.	*Prunus cerasus* L. Rosgill NY 538168
4. Fr*uctu* parvo Serotino. BERNARD-CASTLE and on TEES-bank nigh THORP.	*Prunus cerasus* L. Barnard Castle Thorpe Hall NZ 104141
2. Avium, nigra et racemosa. **S**	*Prunus padus* L.

Cerefolium Sylvestre. **Common wild Chervil. S** ad genua intumescens cicutariae simile. COCKERMOTH-castle	*Chaerophyllum temulentum* L. Cockermouth Castle NY 125310

page 30

Ceterach. Vide Asplenium.

Chamaeacte. Vide Ebulus.

Chamaecistus. **Dwarf Cistus**, or Little **Sunflower**.

1. Vulgaris, flore Luteo. **S**	*Helianthemum chamaecistus* Mill.

42. The Wetheral Safeguards, known also as St. Constantine's Cells (NY 467536), are caves cut into the rock upstream from Wetheral Abbey. There are three chambers, each with its own entrance from a gallery in front and they were, supposedly, places of refuge (*CW1* xv 329–31).
43. On the Tithe map of Great Salkeld Parish, 1840, Crakeld Holm is shown as a field by the river in a position NY 560378. See also C. J. Gordon, "Place and Field Names at Great Salkeld, Cumberland," *CW2* xxv 114–27.
44. The area of the willies (willows) was on the Little Salkeld side of the water between Briggle Beck and the Eden. Of the grounds referred to by Nicolson: Willies grassing, Willies wood and the Willies bridge (pages 66, 88, 12), only the last is preserved as a field name (Bridge Willows) on the 1840 Tithe map of Great Salkeld Parish. A grassing is a field let (sometimes by auction) for grazing.

2. Montanus.
   1. folio pilosellae minoris. On the rocks about KENDAL J.R. CONSWICK-Scar — *Helianthemum canum* (L.) Baumg. Cunswick Scar SD 491940
   2. Polij folio.

Chamaedrys Spuria. **Wild Germander. S** Alpina Cisti flore. — *Veronica chamaedrys* L.

page 31 Chamaemelum, vulgare amarum. **Common Camomile**. Ab Hortensi (seu Romano) distinctum. AS. Argentille chrysanthemum. Vide Buphthalmum. Vide Cotula.

Chamaemorus. **Knot-berry**, or **Knout-berry-bush**. T.L. INGLEBOROUGH, SHAP, CROSS-FELL, etc. — *Rubus chamaemorus* L. Cross Fell summit NY 687343
Saxatilis. Vide Rubus.
Omnino Vide RAIJ Hist. Plant. vol. i. p. 653. 654.

Chamaepericlymenum (male). **S Dwarf Honey-Suckle.** CHEVIOT-Hills. J.R. — *Chamaepericlymenum suecicum* (L.) Aschers & Graebn.

Chamaepytis. **Ground-pine**.

Chelidonium. **Celandine**.
1. Majus vulgare.
2. Minus. **Pilewort. S** — *Ranunculus ficaria* L.

page 32 Chondrilla viscosa humilis. **Dwarf Gum-Succory.**

Christophoriana vulgaris. **Herb Christopher** or **Bane-berries**. MALHAM-COVE in CRAVEN. In the lanes, the land-way,[45] 'twixt — *Actaea spicata* L. Malham Cove SD 897641

45. As opposed to the way over the sands of Morecambe Bay, which was a frequently used route into Furness for those travellers who approached from the south. The road across the moss between Levens and Sampool Bridge was not built until about 1820, and the landway to which Nicolson refers was the road out of Brigsteer which skirted the high ground close in under Whitbarrow Scar and passed through Witherslack to Tow Top Hill and thence to Cartmel. For the various ancient routes into Furness, see John Fell, "The Guides over the Kent and Levens Sands, Morecambe Bay," *CW1* vii 1–26.

KENDALL and CARTMEL. Aug. 9.91. Shaw-wood[46] at Stainbank-green abundantly. — Cartmel SD 381788

Chrysanthemum Segetum. **Corn-Marigold. S** — *Chrysanthemum segetum* L.

Cichoreum. **Succory.** Κιχόριον et Κιχόρειον.[47]
1. Sylvestre, officinarum.
2. Luteum. Vide Hieracium Asperum.

Cicuta. **Hemlock**.
1. Vulgaris major. **S** — *Conium maculatum* L.
2. Minor. Fool's Parsley. **S** — *Aethusa cynapium* L.

page 33

3. Palustris, vel Phellandrium.
4. Sylvestris. **Wild Cicely**, or **Cow-weed**. ULNDALE-parsonage. — *Anthriscus sylvestris* (L.) Hoffm. Uldale NY 250370

Circaea Lutetiana. **Enchanters Nightshade**. By the Church at KIRK-OSWALD. — *Circaea lutetiana* L. Kirkoswald Church NY 555410

Cirsium Anglicum. **Single headed soft Thistle.** SHAP. — *Cirsium dissectum* (L.) Hill
2. Britannicum Clusij repens. Melancholy Thistle.
3. Humile montanum polyanthemum.

Clematis Daphnoides minor. **Periwinkle.**

Clinopodium. **Basil**.
1. Majus. **Great wild Basil. S** — *Clinopodium vulgare* L.
2. Minus. i.e. Acinos Anglica.

Clymenon. Vide Androsaemum.

page 34

Cnicus sativus. Bastard Saffron.

---

46. This wood has, apparently, been cut down long since, but the name is preserved in the field called The Shaws, marked on the Corn Rent Map: "Plan of the Township of Helsington", award 1836. It is in a position SD 502914.
47. The Greek word for succory would be more usually plural in form and spelt with a χ, so *κίχορα* or *κιχόρεια*.

Cochlearia. **Scurvy-grass**.

1. Britannica vulgaris. **Sea-Scurvy-grass**. [On Hartside, Cross-fell, etc. in the mosses]
2. Major Rotundifolia. **Garden-Scurvy-grass**. On HARTSIDE, CROSS-FELL, etc. in the mosses. Vide et J.R. — *Cochlearia officinalis* L. ssp. alpina (Bab.) Hook. Hartside Pass NY 647419 Cross Fell summit NY 687343
3. Minor Rotundifolia.
4. Marina folio anguloso parvo. Fasc. a T.L. Isle of WALNEY. Vix a priore distincta. J.R. — *Cochlearia danica* L.

Colchicum Anglicum Purpureum. **Meadow-Saffron**. In several Closes about KNARESBROUGH and [...] SHERBURN in YORKSHIRE. — *Colchicum autumnale* L.

page 35 Conferva. \1. Linum Aquaticum./ **Hairy Riverweed**, or **Crowsilk. S**

2. Palustris Anglica. Marsh-thread.
3. Reticulata. Fasc.
4. Trichodes. Water Maiden-hair.

Conyza. **Fleabane**.

1. Major. **Plowman's Spikenard**. CUNSWICK-SCAR. Rock at CARTMEL-WELL.[48] — *Inula conyza* DC. Conswick Scar SD 491940
2. Media palustris. **Middle Fleabane**. In Lakes, common. — *Pulicaria dysenterica* (L.) Bernh.
3. Minima. **Dwarf Fleabane**.
4. Coerulea Acris.
5. folijs Laciniatis. PILLIN-MOSS in LANCASHIRE. J.R. — *Senecio palustris* (L.) Hook[49] Pilling Moss SD 415465
6. Helenitis folijs non laciniatis.

Consiligo Plinij. Vide Helleborus niger Hortensis flore viridi.

---

48. See above n 13 page 11.
49. J. A. Wheldon and Albert Wilson, *The Flora of West Lancashire*, (Liverpool, 1907), page 222, cast doubt upon this record but Ray, whose find it is, was acquainted with the plant from the fen land around Ely (*Synopsis 1742*, page 174) and reports this at Pilling Moss as growing in ditches – the classic habitat of *Senecio palustris*. Pilling Moss was drained progressively from the end of the 18th century but, prior to drainage, conditions would certainly have been suitable for *S. palustris*.

page 36

Consolidá. **Comfrey**.

1. Major
   1. flore albo. GRINTON. SWADALE.[50] WENCEDALE. etc. — *Symphytum officinale* L. Grinton SE 047984 Wensleydale
   2. flore purpureo. FLOOK-BARROW. SOWRBY in FOURNESS. — *Symphytum officinale* L. Flookburgh SD 367759 Sowerby Hall SD 198725
2. Media. Vide Bellis major et Bugula.
3. Minor. Vide Bellis minor.
4. Regalis. Vide Delphinium.

Convolvulus. **Bindweed.**

1. Major Albus. Ad sepes, locis Aquosis. **S** — *Calystegia sepium* (L.) R.Br.
2. Minor vulgaris, Arvensis. **S** — *Convolvulus arvensis* L.
3. Niger. **S** Inter. Segetes. — *Polygonum convolvulus* L.
4. Minimus. Fasc.

page 37

Corallina. i.e. Muscus Marinus. **Sea-Coralline**.

1. Ramosa parva. Alba, Offic*inarum*. In the Isles of WALNEY and FOWLEY. — *Corallina* sp[51] Foulney Island SD 247640
2. Pennata longior. Rubens.

Corallina montana fruticosior. Shrubby coralline moss.

Corallium album minus. **Small white Coral**.
ὅτι ἐν ἁλὶ χορεῦται[52]
minimum, calcarijs rupibus adnascens.

---

50. Pronounced to rhyme with "dawdle", this is a thoroughly Cumbrian rendering of "Swaledale" by the Cumbrian Archdeacon.
51. For a note on this and Nicolson's next species, for which he gives no location, see comment by James H. Price in ed. J. S. L. Gilmour, *Thomas Johnson, Botanical Journeys in Kent and Hampstead*, (Pittsburgh, 1972), page 143.
52. i.e. because it dances in the sea. This is perhaps a piece of folk etymology, resting on the similarity between the Greek verb meaning to engage in choral dances (*χορεύω*), and the word for coral (*κοράλλιον*). The etymological speculation is assisted by, if indeed it was not suggested by, the chime in English between *choral* and *coral*. The form *χορεῦται* is an emendation. Nicolson's Greek gives no sense; it can be transcribed *χορεῖλαι*, or perhaps *χουρεῖλαι*, *χουρεῖται* or *χορεῖται*.

Cornu Cervinum. **Harts-horn**, or **Buck-horn Plantaine**. Sea-shore at St BEES-Head. Isle of WALNEY.
Plantago coronopus L.
St Bees Head NX 952133

Cornus foemina. **Dogberry, Gatter-Tree**, or **Female Cornel**. T.L. BRIG-STEER. THRIMBY-Gill.[53] Red brow[54] at little STRICKLAND.
Thelycrania sanguinea (L.) Fourr.
Brigsteer SD 482896

Coronopus. Vide Cornu Cervinum.
Ruellij. Swines-cresses. BIGGER in the Isle of WALNEY.
Coronopus squamatus (Forsk) Aschers
Biggar SD 191662

Corylus Sylvestris. **The Hasel-Nut-Tree. S**
Corylus avellana L.

page 38

Cotula foetida. **Stinking Mayweed** or **Maithes. S** non foetida. seu, Chamaemelum inodorum. **Dog's Camomile. S**
Anthemis cotula L.
Tripleurospermum maritimum (L.) Koch ssp. *inodorum* (L.) Hyl. ex Vaarama.

Cotyledon.

1. Aquatica. **Marsh Pennywort**, or **Whiterot. S** — *Hydrocotyle vulgaris* L.
2. Hirsuta. **Hairy Kidney-wort.** [T.L.] HARDKNOT and WREENOSE. BUCK-BARROW. HARTSIDE. KIRKSTON. — *Saxifraga stellaris* L.
   Hardknott Pass NY 231014
   Wrynose Pass NY 278027
   Buckbarrow Crag NY 482075
   Hartside Pass NY 647419
   Kirkstone Pass NY 401081
3. Vera. Vide Umbilicus Veneris.

Cracca. Vide Aracus.

Crassula. Vide Telephium.

page 39

Crataeogonon. **Wild Cow-wheat**; or **Eye-bright-Cow-wheat. S**
folijs brevibus obtusis Westmorlandicum. T.L. et J.R.
Odontites verna (Bell.) Dum.

---

53. Probably Town Gill, through which the R. Leith flows, downstream from Thrimby Mill (NY 558206).
54. Not located, but probably a field on the steep eastern side of the R. Leith, where the Tithe map of Little Strickland Parish, 1841, names several fields whose names incorporate the word "brow".

Crista Galli. Vide Pedicularis.		
Crithmum, seu Foeniculum marinum. **Sampire**, a Gall*icis* SAINCT PIERRE. 1. Chrysanthemum, sive flore Luteo. q*uo*d penitus a crithmo vero differt. 2. Spinosum, sive pastinaca marina. T.L. in Fasc. at ROOSBECK in Low FOURNEIS.	 Roosebeck SD 258678 ? *Echinophora spinosa* L. (*Watsonia*, viii, 397)	
Crocus verus Sativus. **True or Common Saffron**.		
Cruciata. **Crossewort**, or **Mugweed. S**	*Galium cruciata* (L.) Scop.	
Cuscuta. **Dodder**.		page 40
Cyanus minor vulgaris. **Blewbottle. S**	*Centaurea cyanus* L.	
Cynocrambe. Mas et Foemina. **Dog's Mercury. S**	*Mercurialis perennis* L.	
Cynoglossum. **Hounds-tongue.** 1. Majus vulgare. CARLILE. Inter rudera frequens. 2. Minus, folio virente.	 *Cynoglossum officinale* L.	
Cynosorchis. Vide Orchis.		
Cyperus. **Cyperus-Grass.** 1. Gramineus Milliaceus. **Millet Cyperus-grass**. 2. Longus inodorus Sylvestris. In palustribus. **S** CONSWICK-Tarn. Above and below Common-Holm bridge at great STRICKLAND. 3. Odoratus radice longa. **English Galingale**. Hic est Cyperus officinarum. 4. Rotundus inodorus Anglicus. In Angliae Borealis fluminum littoribus. J.R. \Fasc./ Ita ut hic idem qui – 5. Inodorus Septentrionalium. Johns. **Round Salt-marsh Cyperus**.	 *Cladium mariscus* (L.) Pohl Cunswick Tarn SD 490938 Commonholme Bridge NY 576247 *Scirpus maritimus* L.	page 41

D

Daucus. i.e. Pastinaca Sylvestris. **Wild Carrot**, or **Birdsnest**. On the bank below Langwathby. **S**	*Daucus carota* L. Langwathby NY 570335	page 42

Delphinium. Larks-heels, or Larks-spurs.

Dens Leonis. Caput monachi. Rostrum porcinum. Taraxacon. **Dandelion. S** — *Taraxacum officinale* Weber, sensu lato
1. Vulgaris.
2. Folijs angustioribus.
3. Hirsutus λεπτόκαυλος.[55] i.e. Hieracium caule aphyllo hirsutum.

Dentaria major. **Great Tooth-wort**, or **Lungwort**. [T.L.]. Great STRICKLAND; under the Rocks below water-fall-Bridge. — *Lathraea squamaria* L. Waterfalls Bridge NY 551240

Digitalis. **Foxglove**.
1. Vulgaris purpurea. **S** — *Digitalis purpurea* L.
2. flore Lacteo. T.L. At ULVERSTON Town end:[56] And in a Close call'd Mill-bank at LORTON.[57] — *Digitalis purpurea* L. with a white flower.

page 43

Dipsacus. **Teasell.**
1. Sativus; seu Carduus Fullonum. ROSE, about the ponds.[58] — *Dipsacus fullonum* L. ssp. sativus (L.) Thell Rose Castle NY 371461
2. Sylvestris; Virga pastoris major; seu Labrum veneris.
3. Sylvestris minor.

Dryopteris; seu Filicula montana. **Small Stone-Fern**.
1. Alba; seu Adianthum album folio filicis. TROWGILL. SHAP. — *Cystopteris fragilis* (L.) Bernh. Trough Gill NY 587240
2. Nigra Dodonaei. **Common black maiden-hair**. T.L. **S** — *Asplenium adiantum-nigrum* L.
3. Vera Lobelij; seu Querna. **True Oak-fern**.

---

55. with thin, fine stalk.
56. Shown on the O.S. six-inch map, 1850, as a street continuing south from the end of Queen Street as far as the railway station.
57. Adjoining fields called Great Millbanks and Little Millbanks are shown on the Tithe map of Lorton in the Parish of Brigham, award 1840, in a position NY 165255 opposite a walk mill.
58. These were the castle fish ponds. As far back as the early 15th century, there were stews at Rose. They were reported as being grown over with weeds in the Commonwealth Survey of Rose Castle, 1649, but a successor of Nicolson, Bishop Fleming (1735–1747) was to have the fish pond by the great gate cleaned and dug out and a cascade added. James Wilson, *Rose Castle*, (Carlisle, 1912), pages 17, 87, 234.

4. Tragi. Sive, Ramosa nigris maculis punctata. T.L. **S**	*Thelypteris dryopteris* (L.) Slosson	
Dulcamara. Vide Solanum Lignosum.		

E

Ebulus; sive Sambucus humilis. **Dwarf Elder**; **Danewort**; or **Walwort**. In the Lanes about BROWHAM-CASTLE. ROSE. Vide Camb. Brit. n.e.p.23.[59]	*Sambucus ebulus* L. Brougham Castle NY 537290 Rose Castle NY 371461	page 44
Echium. **Bugloss**.		
1. vulgare. **Vipers Bugloss. S**	*Echium vulgare* L.	
2. marinum. Sea-Bugloss. At PARTON Salt-pans.[60] Vide Plura apud J.R. BIGGER in WALNEY.	*Mertensia maritima* (L.) S.F. Gray Biggar SD 191662	
3. Alterum Dodonaei. **Wall-Bugloss**.		
Elaeagnus Cordi; sive Myrtus Brabantica. **Gaule**; **Sweet willow**, or **Dutch myrtle** [T.L]. WRAGMIRE.	*Myrica gale* L. Wragmire NY 452488	
Elaphoboscum; sive Pastinaca Sylvestris. **Wild parsnep**. Qu. Annon Spondylium.		
Elatine. **Fluellin; or Speedwell**.		page 45
1. Foemina; folio acuminato. **S**	*Kickxia elatine* (L.) Dum.	
3. Mas; folio subrotundo.		
Enula Campana; sive Helenium. **Elecampane**. [**S**] At SLEGILL. CARNFORD in LANCASHIRE.	*Inula helenium* L. Sleagill NY 597192 Carnforth	

59. i.e. the new edition of Camden's Britannia which was published in 1695 under the editorship of Nicolson's friend, Edmund Gibson (see above page xxxv ff.). "After the battle with the Danes in *Swornfield*, a certain shrub sprang up (therefore call'd *Dane-ball* or *Dane-wort*, by others *Dwarf-elder*) which is no where else to be found but there, or transplanted from thence." Tradition had it that the Dwarf Elder growing on the sites of old battles against the invading Danes was nourished by their blood. Camden mentions the tradition again in connection with Essex and Norfolk.
60. Salt pans were, until the late 17th century, the only means of obtaining salt, the first rock salt being discovered in Cheshire in 1670. The pans to which Nicolson refers, would be those at Saltom, just north of Parton (NX 979206). See *CW2* xlii 2.

Equisetum. **Horsetail**. Graece Ιππουρις[61]

1. Arvense; longioribus setis. **S**	*Equisetum arvense* L.
2. foetidum, sub Aqua repens.	
3. Nudum, non ramosum. **S** qu. 'twixt SHAP and ANNA-WELL. Shavegrass.	*Equisetum hyemale* L. Anna Well NY 584127
4. Palustre. Quod vel –	
1. Majus. **S**	*Equisetum fluviatile* L.
2. Minus; brevioribus setis. **S**	*Equisetum palustre* L.
5. Lacustre. folijs mansu arenosis.	
6. Sylvaticum; setis tenuissimis.	

page 46 Erica. **Heath**.

1. Vulgaris	
1. Glabra.	
2. Hirsuta.	
2. Baccifera. **Crake-berries. S**	*Empetrum nigrum* L.
3. Brabantica; sive Belgarum pumila. **Broom-heath. S**	*Erica tetralix* L.
4. Juniperifolia; sive folijs Corios multiflora.	
5. Supina maritima Anglica.	
6. Tenuifolia; Unedonis flore. CLIBBURN-Ling.	*Erica cinerea* L. Cliburn Moss NY 577257

Erigerum. Vide Senecio.

Eruca. **Rocket**.

1. Aquatica.	
2. Marina. Common on all our Shores.	*Cakile maritima* Scop.
3. Sylvestris. Castle-walls at CARLILE.	*Diplotaxis tenuifolia* (L.) DC.
4. Monensis, Laciniata Lutea. J.R. At SELLA-BANK.[62] ISLE OF MAN, WALNEY, etc.	*Rhynchosinapis monensis* (L.) Dandy

page 47 Eryngium. **Eringo**.

1. Vulgare, Mediterraneum. On Friar-goose near NEWCASTLE. T.L.	*Eryngium campestre* L. Friars Goose NZ 277632
2. Marinum. **Sea-Holly**, or **Common Eringo**. All along the Coasts at ALLANBY, WHITEHAVEN, etc.	*Eryngium maritimum* L. Allonby NY 080430

61. decked with a horse-tail.
62. High Sellafield Banks and Low Sellafield Banks lie along the seashore near Calder Hall (NY 033037). These are probably Nicolson's Sella-Bank.

Erysimum. **Hedge-Mustard**. Vide Sophia Chyrurgorum.
1. Vulgare. **S** — *Sisymbrium officinale* (L.) Scop.
2. Latifolium Neapolitanum. about COCKERMOTH-Castle. — *Sisymbrium irio* L. Cockermouth Castle NY 125310

Esula. **Spurge**.
1. Major Germanica. **Water-Spurge**.
2. Exigua. Inter Segetes. On Lance-moor near NEWBY. — *Euphorbia exigua* L. Lansmere NY 575217

Euonymus; Sive Tetragonia. **Spindle-tree** or **Prickwood**. T.L. NATLAND. SKELTON-wood. **S** — *Euonymus europaeus* L. Natland SD 521892 Skelton Wood End NY 406386 — page 48

Eupatorium \Cannabinum./ **Water-hemp**; or **Hemp-Agrimony**.
1. Mas; sive vulgare. **S** River-sides. — *Eupatorium cannabinum* L.
2. Foemina; flore tripartito diviso. **S** cum sequente. — *Bidens tripartita* L.
3. Folio integro. **S** By the Gate at GRASS-BECK.[63] — *Bidens cernua* L.

Euphrasia. **Eyebright**.
1. Vulgaris officinarum. Alba. **S** — *Euphrasia officinalis* agg.
2. Lutea Latifolia palustris.
3. Rubra. Vide Crataeogonon Euph*rosine* facie.
4. Rubra Westmorlandica. ORTON. J.R. — *Bartsia alpina* L. Orton NY 623083
5. Coerulea. i.e. Myosotis Scorpioides.

F

Faba. **A Bean.** — page 49
1. Major Hortensis
    1. Albus.
    2. Ruber.
2. Minor, Equina. **S** — *Vicia faba* L. var. minor equina

---

63. Not located, unless this is Grizebeck (SD 240850). On page 79, Nicolson refers to "the Caus-way from Grass-beck" and the road to Broughton crosses the moss west of Grizebeck by a causeway known as Wreaks Causeway.

Fagus. **Beech-Tree.** T.L.

Farfara. Vide Tussilago.

Fegopyrum; Sive Frumentum Saracenicum. **Buck-wheat**. **S** Sed Sativum. scandens sylvestre. i.e. Convolvulus minor. — *Fagopyrum esculentum* Moench

Ferrum Equinum Germanicum. **Horse-shoe-vetch**. T.L. CUNSWICK-Scar. 'Twixt SHAP and Anna-well. — *Hippocrepis comosa* L. Cunswick Scar SD 491940 Anna Well NY 584127

page 50 Festuca. **Wild oates**.

1. Graminea, glumis Hirsutis.
2. Pilosa; aristis recurvis. Haver. J.R.

Filipendula. **Dropwort**. [T.L.] In the grounds about STAIN-BANK-GREEN. Copiose. CONSWICK-Scar. — *Filipendula vulgaris* Moench Stainbank Green SD 505917 Cunswick Scar SD 491940

Filix. **Fern**.

1. Foemina, vulgaris. **Common brakes**. **S** — *Pteridium aquilinum* (L.) Kuhn
2. Florida; sive Osmunda Regalis T.L. 'Twixt CROOK, and WINANDERMEER. THURSBY-Lane; near the gate leading to CROFTON. Isle of MAN. J.R. MARSH-GRAINGE. Eller-mire[65] at DRUMLENING. — *Osmunda regalis* L. Crook SD 461950 Thursby NY 324502 Crofton NY 304500 Marsh Grange[64] SD 220797
3. Filicula Marina Anglica. At PARTON-Saltpans.[66] — *Asplenium marinum* L.
4. Mas vulgaris, non Ramosa. **S** \and/ Ram*osa* pin*nulis* dentatis. — *Dryopteris Filix-mas* agg.

---

64. See above n 14 page 11.
65. Eller Mire is a meadow near Drumleaning shown on the Tithe map of Aikton Parish, 1843, in a position NY 277519.
66. See above n 60 page 31.
67. This is almost certainly *Cryptogramma crispa* again. Ray's plant, with which Nicolson seems tentatively inclined to equate the Lawson find, is possibly some sort of buckler or shield fern. The entry may have been duplicated because the records on page 7 were accepted from Ray, whereas here Nicolson has obviously seen the plant for himself.

5. Montana florida perelegans.[67] T.L. [Vide Dryopteris...] Qu. An Montana Ramosa minor argute denticulata J.R. p. 27. SKIDDAW. SHAP-Abbey.

Skiddaw summit NY 260290

Shap Abbey NY 548152

page 51

6. Palustris; seu Aquatica.
7. Saxatilis. **Stone-Fern**.
   1. Crispa perelegans. Vide Adianthum album floridum.
   1. Mas, non Ramosa, nigris maculis punctata. **S** — *Thelypteris limbosperma* (All.) H.P. Fuchs
   2. Pedicularis; rubris folijs subtus villosis.
   2. Ramosa Maritima Nostras. Fasc.
   3. Alpina Crispa; caule tenui fragili. **S** Vide Dryopteris Alba. YORKSHIRE, WESTMERLAND, etc. J.R. — *Cystopteris fragilis* (L.) Bernh.
   4. Tragi. Vide Adianthum 'Ακρόστιχον[68] et Dryop*teris*.

Flammula. Vide Ranunculus flam*meus*.

Flos Adonis. **Adonis flower**; or **Red Maithes**.

Foeniculum vulgare. **Common Fennel**. AS fynel. Vide Crithmum. Peucedanum.

Fragaria. **Strawberry**.

page 52

1. Vulgaris. **S** — *Fragaria vesca* L.
2. Minime vesca; seu sterilis.
3. fructu hispido.

Fraxinus vulgaris. **Common Ash-Tree. S** AS. Acse. Aesc. — *Fraxinus excelsior* L.

Fucus; sive Alga marina. **Sea-wrack** Gr*aece* φυκος.[69]

1. Balteiformis. Sea-belt. On the YORKSHIRE-Coasts. I.B. — *Laminaria saccharina* (L.) Lamour.
2. Gallopavonis pennas referens. The **Turkey-feather**.
3. Chordam referens teres praelongus. Sea-Laces.

page 53

3. Gramineus. **Grass-wrack**.
4. Kali geniculato similis, non tamen geniculatus.

---

68. forked
69. seaweed

	Entry	Identification
	5. Latifolius minor.	
	6. Latifolius major, dentatus.	
	7. Membranaceus purpureus parvus. Et Ceranoides: SCOTIS, Dils; HIBERNIS, Dulesh; NORTHUMBRIS, Dulse.	
	8. Phasganoides et polyschides. **Sea-girdle** and **Hangers**. SCOTLAND and the Isle of MAN. J.B.	*Laminaria digitata* (Huds.) Lamour.
	9. Porus Cervinus Imperati.	
	9. Quercus marina. **Sea-thongs**. Vulgatissimus; Latifolius, et Angustifolius.	
	10. Radicibus Arborum fibrosis similis. Fasc.	
	11. Spongiosus nodosus. **Sea-ragged-staff**. 'Twixt MARSH-GRAINGE and the Isle of WALNEY.	prob. *Dumontia incrassata* (O.F. Muller) Lamouroux[70] Marsh Grange[71] SD 220797
page 54	12. Scoticus latissimus edulis dulcis.	
	13. Tenuifolius, folijs dentatis.	
	14. Teretifolius spongiosus parvus.	
	15. Tinctorius. **Dyers-wrack**. About BAMBERG, J.B. et, BRIDLINGTON.	Bamburgh NU 180350
	16. Exiguus – 1. folijs Longiusculis, crassis et subrotundis. 2. Dichotomus Arenacei coloris.	
	17. Vesiculis longis siliquarum emulis. **With pod-like bladders.**	

Fumaria. **Fumitory**.

Entry	Identification
1. Vulgaris purpurea. **S**	*Fumaria officinalis* L.
2. Alba Latifolia, scandens. **S** At STONE-RAISE, below AIKBURN. On Thatch'd Houses and shivery[72] Rocks in the Barony of KENDAL, ubique.	*Corydalis claviculata* (L.) DC. Stone Raise NY 551382 Aikton Castle
3. Major Scandens, flore pallidiore.	

70. So identified by James H. Price in a note in Gilmour, *Thomas Johnson*, pages 141–42.
71. See n 14 page 11.
72. "Apt to split into flakes, brittle, flaky" (*O.E.D.*). The Barony of Kendal covered a large part of South Westmorland.

Fungus. **A Mushrome**. *quasi dicat* Ad funus agens.[73] — page 55

1. Albus Campestris, pediculo brevi et Crasso. **True Champignon. S** — *Agaricus campestris* L. ex Fr.
2. Albus, venenatus viscidus. pediculis longioribus, et latioribus Capitulis.
3. Arborum et Lignorum putrescentium; coloris varij. Malignus.
4. Acetabulorum more cavus, radice carens. Noxius.
5. Arboreus ad Ellychnia. **Touchwood.**[74] **S**
6. Favaginosus. **Honey-comb-Mushrom.** Esculentus. **S** — *Morchella esculenta* Pers. ex St Amans
7. Fraxineus. **Ash-mushrome; hard and dry. S** — *Daldinia concentrica* (Fr.) Cesati & de Notaris — page 56
8. Intybaceus, ad radices Quercuum Endivelike Mushrome.
9. Luteus, se contorquens. In Sylvis, sub Arboribus. Esculentus.
10. Maximus rotundus pulverulentus. Germ*anice* pfofist.
11. Major pediculo brevi crasso, lamellis crebris albentibus. Fasc.
12. Minor tenerrimus farina respersus pileolo superne cinereo, lamellis subtus tenuissimis nigris.
13. Minimus pediculo longo tenuissimo lactescens.
14. Multi ex uno pede, perniciosi. **Yellow Cluster-Mushrom.** — page 57
15. Ophioglossoides; Niger. T.L.
16. Parvus, pediculo tenui longo, pileolo in medio fastigiato, etc. Fasc.
17. parvus Luteus, ad Ophioglossoiden accedens.
18. phalloides; virilis penis effigie. Noxius.
19. Piperatus albus lacteo succo turgens. Esculentus. In CRAVEN. J.R. — *Lactarius piperatus* Scop. ex Fr.

---

73. with which, as might be supposed, is linked the word "funus" (= a funeral). This is, in fact, a piece of spurious etymology.
74. "A name given to various fungi, esp. two species of Polyporus (*P.* or *Fomes fomentarius* and *P.* or *F. igniarius* [i.e. *Phellinus igniarius*], also called *Touchwood Boletus*, or to the tinder called 'amadou' made from them." (O.E.D.).

20. Pratenses parvi, externe viscidi; Albi, Lutei vel Rubentes.
21. Parvus denticulatus.

page 58

22. Parvus Let \h/alis galericulatus. forte Idem cum Num*ero* 16.
23. Porosus magnus Crassus. Qu. An diversus a *deci*mo.
24. Pediculo bulboso, pallidus, maculatus.
25. Pusillus, pileolo tenui utrinque striato.
26. Quercinus, Niger.
27. Ramosus; flavus et albidus. **Coralline Mushrome**.
28. Ramosus Niger compressus parvus.
29. Sambucinus; sive Auricula Judae. **Jews-Ear. S** — *Hirneola auricula-judae* Berk.
30. Calyciformis Seminiferus. Fasc. Plures superaddi possint, cum Arborei tum terrest*re*.

G

page 59

Galega Sylvestris. i.e. Aracus, sive Vicia.

Galeopsis legitima Dioscoridis. **Hedge-Nettle. S** — *Stachys sylvatica* L.

Gallium. **Ladies Bedstraw**.
1. Palustre, Album. **S** — *Galium palustre* L.
2. Luteum, in pascuis siccioribus. Cestrensibus, **Cheese-rening**: Ebor*acensis* **Maiden-hair**. **S** — *Galium verum* L.

Genista. **Broom**.
1. Vulgaris, non Spinosa. Quae et Chamaegenista. **S** — *Sarothamnus scoparius* (L.) Wimmer ex Koch

page 60

2. Spinosa. **Whins**; **Furze**, or **Gorse**. **S**
   1. Vulgaris, major; aculeis longioribus. **S** — *Ulex europaeus* L.
   2. Minor. 'Twixt LOWTHER and HACKTHORP. — *Ulex gallii* Planch. Lowther NY 537237 Hackthorpe NY 544232
   3. Minor [, brevibus aculeis, et pallidius virens.] Aspalathoides. **Needle Furze**; or **Petty whin**. **S** — *Genista anglica* L.

3. Tinctoria Germanica. **Greenweed Diers-weed**, or **Woodwaxen**. **S** Nostratibus, **Wood-wash**. — *Genista tinctoria* L.[75]

Gentiana Concava. Vide Saponaria folio convoluto.

Gentianella. **Dwarf Gentian**, or **Fellwort**.

1. Fugax minor; Centaureae minoris folijs. On HELSFELL-Nab. — *Gentianella amarella* (L.) Börner prob. includes *G. campestris* (L.) Börner Helsfell Nab SD 503936
2. Palustris Angustifolia. In AULSTON-MOOR. — *Gentiana pneumonanthe* L. Alston Moor
3. Flore Lacteo. T.L. about Great STRICKLAND. — prob. *Gentiana pneumonanthe* Great Strickland NY 560230

Geranium. **Cranes-Bill**.

page 61

1. Batrachoides. **Crow-foot-Cranes-Bill**.
   1. Flore Coeruleo, in pratis. **S** — *Geranium pratense* L.
   2. Montanum \nostras/; flore Aconiti. On BEETHAM-bank. etc. MORLAND. — *Geranium sylvaticum* L. Beetham Bank SD 513973 Morland NY 600225
   3. flore eleganter variegato.[76] T.L. in old DEER-PARK by THORN-THWAIT.[77]

---

75. This plant was originally of great importance to the Kendal woollen industry in the production of the famous Kendal green. Cornelius Nicholson, *The Annals of Kendal*, 2nd ed. (1861), describes it on pages 238–9: "A plant which is known to have abounded in the neighbourhood of Kendal many years ago, though it be now nearly uprooted, called by Linnaeus *genista tinctoria*, and commonly called "Dyer's Broom," was brought in large quantities to Kendal, from the neighbouring commons and marshes, and sold to the dyers. The plant, after being dried, was boiled for the colouring matter it contained, which was a beautiful yellow. The cloth was first boiled in alum water, for the mordant, and then immersed in the yellow dye. It was then dried, and submerged in a blue liquor extracted from *woad*."
76. This might be a variety of either *Geranium pratense* or *Geranium sylvaticum*. Ray, *Synopsis 1724*, pages 360–61, enters it with *Geranium pratense* ("Geranium batrachoides" no. 17) but C. C. Babington, *The Correspondence of John Ray*, ed. Edwin Lankester, (1848), page 202 identified it as *G. sylvaticum*.
77. Thornthwaite Hall NY 513163 had been a hunting seat of the Curwens who had sold it to Lord William Howard of Naworth (Joseph Whiteside, *Shappe in Bygone Days*, (Kendal, 1904), page 89). When Nicolson provided Thomas Robinson with the record, he refers to the location as "Mr. Howard's Park at Thornthwait" (See page xlii).

2. Columbinum. **Doves-foot-Cranes-Bill**.
   1. Maximum; folijs Dissectis. Pl. Oxf.
   2. Vulgare; folio Malvae rotundo. **S**

      Geranium molle L.
   3. Pediculis florum longissimis. **S**

      Geranium columbinum L.
   4. Majus, flore minore coeruleo.
3. Haematodes. **Bloody Cranes-Bill**. [T.L.] Lancastrense a vulgari specie differt. Fasc. WALNEY. St. Bees-Head. Malham-Cove.

   Geranium sanguineum L. *var. lancastrense* (With.) Druce
   St Bees Head NX 952133
   Malham Cove SD 897641

page 62

4. Moschatum, cicutae folio. **Musked Cranesbill**, or **Muscovy**. in CRAVEN.

   Erodium moschatum (L.) L'Hérit.[78]
5. Moschatum inodorum; in Arenosis siccioribus. **S**

   Erodium cicutarium agg.
6. Inodorum Album, cicutae folio. TUMBERHIL in the Isle of WALNEY. GRINTON.

   Erodium cicutarium agg.
   Tummer Hill SD 179675
   Grinton SE 047984
7. Maritimum, Pusillum, Betonicae folio.
8. Robertianum; flore purpureo vel albo. **Herb Robert**. **S** [...]

   Geranium robertianum L.
9. Saxatile Lucidum. **Stone-Doves-foot**.**S**

   Geranium lucidum L.

Gladiolus Palustris, seu Lacustris. flore subcoeruleo. **Water-gladiole, or the flowering Rush**. T.L. HULS-WATER. WINANDERMEER. GRAYSON-TARN[79] by COCKERM*ou* \TH/ In TARN-WADLING,[80] near HESKETT. flowers, etc. within the water.

Lobelia dortmanna L.
Ullswater

78. This is a Raian record and Lees, *Flora of West Yorkshire*, pages 182–83 was doubtful of it, wondering whether, perhaps "some luxuriant form of *E. cicutarium* was not the plant really found."
79. Nicolson's spelling follows local pronunciation of the name Greysouthen. According to local opinion, the tarn lay half a mile north of the village, in a hollow (NY 074296) south of the house called Tarn Bank. It has been drained and is now given over to agriculture.
80. This tarn, which is noticed as early as the 14th century on the Gough map, was drained at the beginning of the 19th century. By mid-century, however, it had reverted to water: the 1845 Tithe map of Brownrigg and Lazonby Parish shows it as 21 acres of open water, and it was redrained about mid-century (Collingwood, *The Lake Counties*, page 147). This was again not entirely successful, further drainage work being done during the 1939–45 war by Italian prisoners. The position is NY 485445.

Glastum Sativum. **Woad**. Caesar, Comment. de Bell. Gall. lib. 5.[81]		page 63
Glaux. **Milk-wort**.		
1. Hispanica, Sylvestris. Near HUNTCLIFF-rock in CLEEVELAND.	Hunt Cliff NZ 695218	
2. Maritima. **Sea-Milkwort**, or **Black Salt-wort**. CARTMEL-Grainge.[82]	*Glaux maritima* L.	
3. Vulgaris Leguminosa. i.e. Glycyrrhiza Sylvestris. In a small Island at the head of WALNEY.[83]	*Astragalus glycyphyllos* L.	
Glycyrrhisa. **Liquorice**.		
1. Vulgaris, Siliquosa. planted for sale about PONTEFRAICT. J.R.	*Glycyrrhiza glabra* L.	
2. Sylvestris. **Wild Liquorice**, or **Liquorice-vetch**. Vide Glaux vulgaris.		
Gnaphalium. **Cudweed**; or **Cotton-weed**.		page 64
1. Vulgare. Filago. Herba impia. **S** Nostras CHAFEWEED.	*Filago germanica* (L.) L.	
2. Anglicum; folio longiore. In Arenosis. **S**	*Gnaphalium sylvaticum* L.	
3. Maritimum.		
4. Medium; vulgari simile. **S**	*Gnaphalium uliginosum* L.	
5. Minimum repens. KENDALL-fell.	*Filago minima* (Sm.) Pers. Kendal fell summit SD 504930	
6. Montanum Album. **S**	*Antennaria dioica* (L.) Gaertn.	
Gramen. **Grass**.		
1. Agrorum Arundinaceum. **Corn-Reed-Grass**, or **Bent**.		
2. Alopecuroides. **Fox-tail-grass**. Majus et Minus.		page 65
3. Aquaticum fluitans multiplici spica. **Flote-grass**.		

81. "All the Britons, indeed, dye themselves with woad, which produces a blue colour, and makes their appearance in battle more terrible." Caesar, *The Gallic War*, trans. H. J. Edwards, (1917), page 253.
82. See above n 3 page 6.
83. No such small island exists today at the northern end of Walney.

4. Arundinaceum Aquaticum, panicula Avenacea.
5. Arundinaceum acerosa gluma nostras. **Great Reed-grass with chaffy heads**.
6. Asperum, spicatum pratense. **Rough-grass**. **S** — *Dactylis glomerata* L.
7. Avenaceum. **Oat-grass**.
   1. Capillaceum, minoribus glumis. **Red oat-grass of the woods**.
   2. Dumetorum spica simplici. **S** — *Zerna ramosa* (Huds.) Lindm.

page 66

   3. Dumetorum panicula Sparsa.
   4. Montanum; spica simplici, aristis recurvis.
   5. Spica gracili, aristis brevibus recurvis.
   6. Paniculis non Aristatis. Small **Oat-grass without Awns**. **S** — *Sieglingia decumbens* (L.) Bernh.
   7. panicula flavescente. **S** — *Trisetum flavescens* (L.) Beauv.
   8. panicula purpuro-Argentea splendente.
   9. Panicula e Spicis raris strigosis composita, aristis tenuissimis.
8. Caninum. **Dogs-grass**.
   1. Arvense, repens. **Common Quich**, or **Couch-grass**.

page 67

   2. Nodosum, vulgare. **Knotted Dogs-grass**.
   3. Maritimum, spica foliocea.
   4. Maritimum paniculatum.
   5. Supinum paniculatum dulce.
   6. Paniculatum Molle. Fasc.
9. Capillaceum, locustellis pennatis non Aristatis.
10. Cristatum Anglicum, laeve. **Smooth crested Grass**. **S** — *Cynosurus cristatus* L.
11. Cyperoides. **Cyperus-grass**.
   1. Majus Latifolium; cum caule triangulo. In fossis. **S** — *Carex riparia* Curt.
   2. Angustifolium. Luteo-Nigrum, **S**

page 68

3. Ex monte Ballon.[84] T.L. **S**
4. Polystachion lanuginosum.
5. Polystachion spicis parvis longissime distantibus.
6. Palustre. Majus et Minus.
7. Pulicare. Flea-grass. 'Twixt BANISTER-HEAD and WATER-SLEDDAL. — *Carex pulicaris* L. Bannisdale Head NY 515043 Wet Sleddale
8. Spicatum. Spicis brevibus congestis, folio molli.
9. Vernum minimum. Three brethren Tarn in WHINFIELD.[85] — *Carex caryophyllea* Latour.

12. Dactylon; seu Galli Crus. Ischaemon vulgare. **Cocks-foot-grass**.
13. Dactyloides repens.
14. Exile duriusculum, in aridis et muris.

page 69

15. Exile duriusculum maritimum.
16. Exile Hirsutum nemorosum.
17. Hirsutum nemorosum majus. **S** — *Luzula pilosa* (L.) Willd.
18. Nemorosum Hirsutum vulgare. In montosis.
19. Junceum. **Rush-grass**.
    1. Capsulis triangulis minimum.
    2. Aquaticum, sparsa panicula.
    3. Leucanthemum. **Bogg-Rush-grass with white chaffy heads**. All the mosses of the North. J.R. — *Rhynchospora alba* (L.) Vahl.
    4. Parvum, seu Bufonium Flandrorum. **Toad-grass**. **S** — *Juncus bufonius* L.
    5. Maritimum. Cum pericarpijs rotundis.

page 70

20. Murorum, Spica longissima. **Capons-tail-grass**.
21. Palustre Echinatum. **Hedge-Hog-grass**.
22. Paniceum. **Panick-grass**.
    1. Spica simplici.

---

84. This plant, recorded from coastal stations in Kent and Essex in Ray's *Synopsis 1724*, page 423 was identified by J. E. Smith, *The English Flora*, (1824–28), vol iv, page 87, as *Carex divisa*. This, however, is unlikely in this location and it is probable that Nicolson made a mistake in identifying some other *carex* sp.
85. On the O.S. six-inch map, Three Brother's Barn in Whinfell Forest (*PNW* ii 132) is shown in a position NY 585273 with Three Brother's Oaks the woodland to the south-west of the barn. No tarn is shown.

	2. Spica divisa.	
	3. Spica aspera.	
	23. Parnassi T.L. **S**	*Parnassia palustris* L.
	24. Parvum Marinum, spica loliacea. **Dwarf-Darnel-grass**.	
	25. Pratense Paniculatum minus. Album et Rubrum.	
page 71	26. Paniculatum mar⟨i⟩timum vulgatissimum.	
	27. pratense panicula sparsa versus unam partem, duriore.	
	28. pratense paniculatum molle. **S** Nostras Dart-grass.	*Holcus lanatus* L.
	29. Parvum praecox, spina laxa canescente.	
	30. Paniculatum Elatius, paniculis squamosis.	
	31. paniculatum bromoides minus paniculis Aristatis, unam partem spectantibus.	
	32. Pumilum Hirsutum, spica purpuro-argentea molli.	
page 72	33. Pumilum Loliaceo Simile. **Dwarf Darnel-like Grass**.	
	34. Secalinum. **Rie-grass**.	
	35. Serotinum Arvense spica laxa pyramidali.	
	36. Spicatum montanum Asperum.	
	37. [Sparteum capite bifido vel gemino]. T.L. in Fasc. Vide Synop. p. 203	
	38. Sparteum Marinum Nostras. **English Sea-Matweed** or **Marram**.	
	39. Sparteum juncifolium. **Small Matweed**.	
	40. Sorghinum. **Sorgh-grass**. JOHNS.	
page 73	41. Sylvaticum Majus. **Greater Wood-grass**.	
	42. Sylvaticum *terti*um Tabernamontani. T.L. **S**	
	43. Tomentosum, et Linogrostis. **Cotton-grass**. **S** Nostras Cats-tail.	*Eriophorum angustifolium* Honck.
	44. Triglochin Dalechampij. **Arrow-headed Grass**. T.L.	*Triglochin palustris* L.
	45. Tremulum majus, sive Phalaris	*Briza media* L.

pratensis. **Quakeing grass**; **Cow-Quakes**, or **Ladies-Hair**. **S**

46. Tiphinum. **Cats-tail-grass**.
    1. Majus.
    2. Minus.
    3. Maritimum minus.

page 74

Graminifolia repens palustris, vasculis granorum piperis aemulis. **Pepper-grass**.

Gratiola Angustifolia. Quae et Hyssopifolia Aquatica. **Small Hedge Hyssop**. Nigh the Church at WINANDERMEER; and by the water side. — *Lythrum hyssopifolia* L.

Gratiola latifolia coerulea. Vide Lysimachia galericulata.

H

Halimus, sive Portulaca Marina. **Common sea-purslane**. HINDPULL by the Isle of WALNEY. ROOSEBECK. — *Halimione portulacoides* (L.) Aell.
Hindpool SD 193696
Roosebeck SD 258678

page 75

Hedera. **Ivy**.
1. Communis Arborea, scandens, corymbosa. **S** — *Hedera helix* L.
2. Helix, sive Repens. A priori vix differt specie; sed loco.
3. Terrestris. **Ground-Ivy**; **Alehoof**; **Gill-go-by-ground**; **Tunhoof**. **S** — *Glechoma hederacea* L.
4. Aquatica. **Water-Ivy**; or **Ivy-leav'd Ducks-meat**. **S** — *Lemna trisulca* L.

Helenium. Vide Enula Campana.

Helianthemum minus. Vide Chamaecistus.

Helleborine. **Hellebore**.

page 76

1. Palustris Nostras. **Bastard white Hellebore**, or **Marsh-Hellebore**. T.L. In a wet meadow under BETHAM-bank; or BARROW-bank near BRIGSTEER. **S** — *Epipactis palustris* (L.) Crantz
Beetham Bank SD 513973
Barrow-bank not located

2. Minor Alba. T.L. et Fasc. in LOWTHER-woods, over against ASKHAM-Hall.	*Cephalanthera damasonium* (Mill.) Druce Askham Hall NY 516239
3. Latifolia montana. In the wood under HELSFELL-nab.	*Epipactis helleborine* (L.) Crantz Helsfell Nab SD 503936
4. Flore Atro-Rubente. CUNSWICK-scar. MALHAM.	*Epipactis atrorubens* (Hoffm.) Schult. Cunswick Scar SD 491940 Malham Cove SD 897641
5. folijs angustis praelongis acutis. Under BRACKENBROW near INGLETON.	*Cephalanthera longifolia* (L.) Fritsch Ingleton SD 695732
6. Flore rotundo. Vide Calceolus Mariae	
Helleborus niger hortensis flore viridi. **Wild black Hellebore**, or **Bearsfoot**. HELSTON-Lathes. Sylvestris Maximus, foetidus.	*Helleborus viridis* L. Helsington Laithes[86] SD 505906
page 77 Helxine. Vide Parietaria.	
Hemionitis. pumila trifolia vel quinquefolia maritima. Mules-fern.	
Hepatica. Vide Lichen.	
Heptaphyllon. Vide Tormentilla.	
Herba Gerardi. Vide Angelica Sylv*estris* minor.	
Herba Paris. Solanum monococcum, tetraphyllon. **Herb-Paris**; **Truelove**, or **One-berry**. T.L.	*Paris quadrifolia* L.
Herniaria glabra. **Common smo⟨o⟩th Rupture-wort**.	

86. Helston Lathes or Helsington Laithes (*PNW* i 123) was the manor house of the manor of Helsington. At this time, it was in the possession of Colonel James Grahme, the Jacobite who bought the manor of Levens and Levens Hall in 1688 from the Bellinghams (R. S. Boumphrey, C. Roy Hudleston and J. Hughes, *An Armorial for Westmorland and Lonsdale*, (Kendal, 1975), page 139).

Hieracium.[87] **Hawkweed**.

1. Asperum; flore majore luteo. **Yellow Succory**.

page 78

2. Caule Aphyllo hirsutum. **Dandelion-Hawkweed**.
3. Echiodes capitulis cardui Benedicti. Vide Buglossum luteum.
3. Fruticosum. **Bushy Hawkweed** (Pulmonaria dictum.)
   1. Latifolium Hirsutum.
   2. Latifolium Glabrum. HULSWATER. J.R. — Ullswater
   3. Angustifolium majus.
4. λεπτόκαυλον[88] Hirsutum folio rutundiori.[89] T.L. 'Twixt SHAP and ANNA-WELL. — Anna Well NY 584127
4. Luteum, Chondrillae folio glabrum. **Smooth Succory-Hawkweed**.
5. Minus, praemorsa radice. **Yellow Devils-Bitt**.
6. Minimum Clusij. **Small Swines Succory**.

page 79

7. Macrocaulon Hirsutum folio rotundiore. T.L. in Fasc. On the banks of EDEN, in March 1691. ni fallor. EDINBURG-PARK,[90] etc. J.R.
8. Maximum Asperum, Chondrillae folio. In the Corn near MALHAM-COVE. — *Crepis biennis* L. Malham Cove SD 897641
9. Montanum Latifolium glabrum minus. **S**
10. Montanum Augustifolium, nonnihil incanum.
10. Montanum non Ramosum, caule Aphyllo, flore pallidiore.
11. Parvum in Arenosis nascens seminum pappis densius radiatis.
12. Radicatum Longius, folio Dentis Leonis obtuso.

---

87. No attempt has been made to identify the various species of Hieracium given below by Nicolson, as the synonyms are much entangled, the identification uncertain and the application unreliable.
88. fine-stemmed.
89. for rutundiore.
90. See above n 6 page 7.

	13. Stoebes folio Hirsutum, odore Castorei. Vide Sonchus. Pulmonaria.	
page 80	Hipposelinum, sive Smyrnium vulgare. **Alexanders**. About SCARBOROUGH-Castle. J.R.	*Smyrnium olusatrum* L. Scarborough Castle TA 049893
	Hippuris. Vide Equisetum.	
	Holosteum. **Stitchwort**.	
	1. Minimum tetrapetalon. **S**	*Moenchia erecta* (L.) Gaertn., Mey. & Scherb.
	2. Minimum palustre, capitulis longissimis filamentis donatis. By the highest Tarn above the Tenters at KENDALL.[91]	*Littorella uniflora* (L.) Aschers.
	3. Palustre repens folijs, capitulis et seminibus Psylli.	
	4. Pumilum non Descriptum. Johns.	
	5. Ruellij Holostei diversitas. **S**	*Stellaria graminea* L.
	6. Vernum, flore majore.	
page 81	Hordeum. Barley.	
	1. Distichum. **Common Barley. S**	*Hordeum distichon* L.
	2. Distichum minus. **Sprat- (or Battledoor-) Barley.**	
	3. Polystichum. **Bear-Barley**, or **Big. S**	*Hordeum vulgare* L.
	4. Spontaneum Spurium. **Wall-Barley**; **Way, Bennet**; **Wild Rye**; or **Rye-grass.** Gramen secalinum. Johns.	
	Horminum Sylvestre. **Oculus** *Chris***ti officin***arum*. **Wild Clary.** Lavendulae flore. Clus. Idem quod prius. J.R.	
page 82	Hyacinthus. **Hyacinth**, or **Harebells.**	
	1. Anglicus, Belgicus vel Hispanicus; flore coeruleo oblongo. **S**	*Endymion non-scriptus* (L.) Garcke
	2. Autumnalis minor. **S**	*Scilla autumnalis* L.
	3. Stellatus vulgaris; bifolius et Trifolius.	
	Hydropiper. Vide Persicaria.	

91. There is nothing very obvious to fit this description – possibly the tarn has been drained long since. For a note on the tenters, see above n 29 page 16.

Hyoscyamus niger vulgaris. **Henbane.** Jusquiamus, officinarum. **S**	*Hyoscyamus niger* L.	
Hypericum. **St John's wort.** Vide Androsaemum. Ascyrum.		
1. Majus; sive Androsaemum Hypericoides. **Tutsan St John's wort.** T.L. at Great STRICKLAND.	*Hypericum hirsutum* L. Great Strickland NY 560230	page 83
2. Elegantissimum non Ramosum, folio lato. Qu. An Androsaemum Campoclarense etc. apud T.L. **S** TROWGILL near CLIBBURN. CONSWICK-Scar.	*Hypericum montanum* L. Trough Gill NY 587240 Cunswick Scar SD 491940	
3. Minus supinum, glabrum. **Creeping St John's wort.** 'Twixt NEWTON and PENNY-bridge.	*Hypericum humifusum* L. High Newton SD 401828 Penny Bridge SD 310830	
4. Minus erectum. Pulchrum Tragi. T.L. at TROWGILL near Clibburn. **S** in most sandy pastures.	*Hypericum pulchrum* L. Trough Gill NY 587240	
5. Vulgare. Perforata. Fuga Daemonum. **S**	*Hypericum perforatum* L.	
Hyssopifolia. Vide Gratiola.		
J		page 84
Jacea nigra vulgaris. **Knapweed**, or **Matfellon. S**	*Centaurea nigra* L.	
Scabiosa, flore purpureo. **S**	*Centaurea scabiosa* L.	
Jacobaea. **Ragwort**, or **Seggrum.**		
1. Major, vulgaris. **S**	*Senecio jacobaea* L.	
2. Minor; Muralis Belgarum. Johns.		
3. Montana; Lanuginosa, Angustifolia non Laciniata.		
4. Palustris Latifolia. In the watery places about CLIBBURN-bridge.	*Senecio aquaticus* Hill. Cliburn Bridge NY 588245	
Illecebra. Vide Sedum minimum.		page 85
Irio. Vide Erysimum.		
Iris palustris lutea. Vide Acorus.		
Isatis. Vide Glastum.		
Juglans, sive Regia nux vulgaris. **The Wallnut-Tree.**		

Juncus. **A Rush.**

1. Acutus vulgaris. Our Mossers[92] and Dalesmen thatch houses with it.	*Juncus inflexus* L.
2. Acutus Cambro-Britannicus. **Moss-Rush**; or **Goose-corn**. Nostras Bent. SHAP-fell.	*Juncus squarrosus* L.
3. Acutus Maritimus Anglicus.	

page 86

4. Acutus maritimus caule triangulo.	
5. Acutus Maritimus capitulis Sorghi.	
6. Acutus Maritimus capitulis rotundis.	
7. Alpinus cum cauda leporina. **S** **Moss-crops.**[93]	*Eriophorum vaginatum* L.
8. Aquaticus maximus. **Bull-Rush.** WINANDER-MEER.	*Scirpus lacustris* (L.) Palla
9. Aquaticus minor; capitulis Equiseti.	
10. Floridus. On the banks of CALDEW, by CARLILE. **S**	*Butomus umbellatus* L.
11. Laevis, panicula sparsa. Seeves.	
12. Laevis, glomerato flore; panicula non sparsa.	

page 87

13. Laevis, panicula glomerata nigricante.	
14. Parvus montanus, cum parvis capitulis Luteis. Nostras Lank. On CLIBBURN-Ling, etc.	*Trichophorum cespitosum* (L.) Hartman Cliburn Moss NY 577257
15. Parvus Scapo supra paniculam longius producto. Fasc. in Westmor*land.*	*Juncus filiformis* L.
16. Juncellus omnium minimus. Capitulis Equiseti. Pl. Oxf.	

Juniperus. **Juniper.**

1. Alpina Minor. **Savine**, in WALES and WESTMORLAND. J.R. By MICKLE FORCE in TEESDALE.	*Juniperus communis* L. *ssp. nana* Syme High Force NY 880284

92. Joseph Wright, *The English Dialect Dictionary*, (London, Oxford and NY, 1903), defines a mosser as one employed in thatching with moss and comments that it was the practice deliberately to put moss under or between slates.
93. Thomas Lawson, writing in 1688, to give Ray a list of northern plants, comments: "After it turns white sheep are greedy after it; so it is called Moss-crops about Clibburn, Water Sledale, and in all places here – Westmoreland" ed. Edwin Lankester, *Correspondence of John Ray*, (1848), page 204. Plant collectors of the day often whimsically styled themselves "moss croppers". "The top of all the moss-croppers," says William Vernon, "is Mr. Buddle, who is a great help to us" (Letter to Richard Richardson, 12th February 1703, ed Dawson Turner, *Extracts from the Literary and Scientific Correspondence of Richard Richardson*, (Yarmouth, 1835), page 73).

2. Vulgaris; baccis parvis purpureis. FOURNESS. So tall as to make hunting-polls.[94] — *Juniperus communis* L.

K

page 88

Kali. **Glasswort**, or **Saltwort.**

1. Geniculatum majus; sive Salicornia. On the sand nigh FLOOKBARROW; copiose. — *Salicornia europaea* L. Flookburgh SD 367759
2. Geniculatum lignosum perenne procumbens.
3. Minus Album.
4. Spinosum Cochleatum; seu Tragus Spinosus. Sea-Grape. On the sands by RAMPSIDE in LOW-FOURNESS. — *Salsola kali* L. Rampside SD 240663

Keiri Arabum, et Officinarum. Vide Leucoium luteum.

Knawell, Germanorum. Vide Polygonum Angustifolium.

L

page 89

Labrum Veneris. Vide Dipsacus.

Lactuca. **Lettuce.**

1. Agnina; seu Locusta herba. T.L. **S** — *Valerianella locusta* (L.) Betcke
2. Folijs serratis.
2. Marina. Vide Lichen marinus.
3. Sylvestris major, odore Opij.
4. Sylv*estris* folijs dissectis.
5. Endiviaefolio \non/ laciniato. Vide Sonchus

94. One of the most popular forms of hunting in Nicolson's day was otter hunting, and poles for this were traditionally very long. The Moses Browne edition of Walton's *Compleat Angler*, (1750), shows otter poles in use in an illustration opposite page 33. The modern hunting pole has, however, been much shortened to fit into the motor car.

	Ladanum Segetum, Angustifolium. **Narrow-leav'd All-heal**, or **Iron-wort.**	
page 90	Lagopus, vel Lagopodium. **Haresfoot**, or **Hares-foot-Trefoil. S**	*Trifolium arvense* L.
	Lamium. **Archangel**; or **De\a/d-Nettle.**	
	1. Album, vulgare. **S**	*Lamium album* L.
	2. Luteum.	
	3. Rubrum; folio subrotundo. **S**	*Lamium purpureum* L.
	4. Rubrum minus, folijs profunde incisis.	
	5. Montanum, melissae folio.	
	6. folio caulem ambiente. **Majus** et Minus. Nigh CLIBBURN-CHURCH. **S**	*Lamium amplexicaule* L. Cliburn Church NY 588245
page 91	Lampsana. **Dog-Cresses**; **Tetter-wort**, or **Nipplewort. S**	*Lapsana communis* L.
	Lapathum. **A Dock.**	
	1. Acutum majus. **S**	*Rumex conglomeratus* Murr.
	2. Acutum minus, folio Crispo. **S**	*Rumex crispus* L.
	3. Acutum minimum.	
	4. Latifolium obtusum. **S**	*Rumex obtusifolius* L.
	5. maximum Aquaticum. **S** MUNTINGIJ Britannica Antiquorum vera.	*Rumex hydrolapathum* Huds.
	6. flore Aureo.	
	7. Folio acuto rubente.	
	7. Pulchrum Bononiense sinuatum.[95] T.L. at New-LOWTHER.	Newtown NY 527241
	⟨8⟩ Acetosum [Rotundifolia] Vide Acetosa.	
page 92	Lappa major. Vide Bardana.	

95. C. C. Babington in *The Correspondence of John Ray*, ed Lankester, page 204 identified this as *Rumex pulcher*, but this seems unlikely in this location as Baker, *Flora*, page 181, observes: "A plant gathered by Lawson 'between the inn and smithy at Sir John Lowther's Newtown' is referred to this species by Professor Babington, but I suspect something else was really intended." For Sir John Lowther's improvements in the 1680's, see Nicolson and Burn, vol i, page 440.

Lathyrus.

1. Major Latifolius.\flore purp*ureo*/**Pease everlasting.** On the rocks at Redneese near WHITEHAVEN. J.R. Item flore rubente. — *Lathyrus sylvestris* L.[96] Redness Point NX 973194
2. Luteus Sylvestris Dumetorum. **Tare everlasting.** At ⟨P⟩ARTON-Salt-pans.[97] **S** Et passim, ad sepes, etc. — *Lathyrus pratensis* L.
3. Siliqua Hirsuta. **Rough-codded Chichling.**

Vide Vicia.

Laureola. **Spurge Laurel**; **Lowry.**

Lens.

1. Minor vulgaris. **Lentils**; or **Dills.**
2. Palustris. **Ducksmeat. S** — *Lemna minor* L.

Lenticula Aquatica Trisulca. Vide Hederula Aquatica.

page 93

Lepidium Latifolium. **Dittander**; or **Pepperwort.**

Leucoium. **Wall-flower.**

1. Luteum, vulgare. Quod et Keiri Arabum. **S** On the walls of the Castle and Abbey at CARLILE,[98] Copiose. — *Cheiranthus cheiri* L.
2. Maritimum magnum, Latifolium. **Broad-leav'd Sea-Stock-July-flower.** ib*idem.* — *Matthiola sinuata* (L.) R.Br.[99]

---

96. There seems to be some confusion here between *Lathyrus sylvestris* and *L. latifolius.* John Ray's "Lathyrus major latifolius" (*Synopsis 1724*, page 319) is *L. latifolius*, but it seems more likely that the Lawson record from Redness entered there is *L. sylvestris.* Nicolson, in writing "Item flore rubente" seems uncertain whether there is one species here or two, but there seems little doubt that his own record was *L. sylvestris.* (See also Baker, *Flora*, page 74).
97. See above n 60 page 31.
98. See above n 37 page 21.
99. There can be no doubt, from the synonyms, that this is the plant that Nicolson intended. It was recorded as far north as North Wales in his day where it was found by John Ray (*Synopsis 1724*, page 291).

3. Vasculo sublongo intorto. INGLEBOROUGH and HINCKLEHAUGH.[1] J.R. — *Draba incana* L.

Lichen. **Liverwort.** qui Lichenas sanat.[2]
1. Fontana, vulgaris. **S**
2. Arborum; seu Pulmonaria. **Tree-Lungwort.**

page 94
3. Cinereus Terrestris. **S**
4. Marinus.
   ⟨1.⟩ \Lactuca marina, dictus./**Oyster-green.**
   [5. marinus] ⟨2.⟩ Tubulosus. Fasc.
   [6. marinus] ⟨3.⟩ folijs laciniatis, minimus.
7. Lunulatus.
8. Minimus, folijs Laciniatis.
8. Petraeus.
   1. Cauliculo Calceato.
   2. Stellatus. Near the foot-bridge by STRICKLAND-Mill. — *Marchantia polymorpha* Strickland Mill NY 551225
   3. Purpureus Derbiensis. In Wett SLEDDALE, copiose.
   4. Umbellatus.

Ligustrum vulgare. **Privet**; **or Prim-print.** DUNERHOLM; near WALNEY. On METHAP Scarr,[3] in West*morland*. — *Ligustrum vulgare* L. Dunnerholme SD 212798

page 95
Lilium Convallium. **Lily of the Valleys**; or **May-Lily.**
1. Album, vulgare. About Water-fall-bridge near Great STRICKLAND. At — *Convallaria majalis* L. Waterfalls Bridge NY 551240

---

1. "Ray ... describes Hincklehaugh as 'overhanging the town of Settle' and as 'about three miles east' of it. This suggests that it is the gritstone eminence on one of the lines of the Craven fault now called Ryeloaf and 1794 feet high; and enquiries establish that this hill was formerly called Inglehow." (Charles E. Raven, *John Ray*, (Cambridge, 1942), pages 160–61). The grid reference for Rye Loaf Hill is SD 857631.
2. Which cures ringworm.
3. The limestone scar about a quarter of a mile south of Low Meathop village at SD 433792 is still an interesting botanical site.

HELSTON.[4] HELSFELL-WOOD. — Helsington Laithes SD 505906; Helsfell Nab SD 503936

2. Angustifolium. T.L. cum priore. — *Convallaria majalis* L.

Limonium Maritimum majus. **Sea-Lavander.** 'Twixt CARTMEL-well[5] and GRAINGE. — *Limonium vulgare* Mill. Grange-over-Sands.

Linaria. **Toad-flax.**
1. Lutea vulgaris. **S** — *Linaria vulgaris* Mill.
2. Montana alba, Adulterina.
3. Odorata monspessulana.

Vide Antirrhinum et Elatine.

Lingua Cervina, officinarum. Vide Phyllitis.

Linum **Flax**. page 96
1. Sativum, vulgare. **S** — *Linum usitatissimum* L.
2. Sylvestre. Quod vix a sativo differt specie. J.R.
   1. Angustifolium.
      1. Floribus dilute purpurascentibus, vel carneis.
      2. Floribus coerulis.
   2. Coeruleum perenne nostras. T.L. \CROSBY-Ravens*worth* and SHAP./**S** About 200 yards above the ford at LADY-FITTS.[6] — *Linum anglicum* Mill. Crosby Ravensworth NY 621148
   3. Catharticum. **Dwarf-wild-flax**; or **Mill-mountain. S** — *Linum catharticum* L.

Lithospermum; sive Milium Solis. **Gromil** or **Gromwel. S**

---

4. See above n 86 page 46.
5. See above n 13 page 11.
6. William Hodgson, who quotes this record in his *Flora of Cumberland*, (Carlisle, 1898), page 67, says that Lady Fitts is on the Eden near Great Salkeld, but I have not been able to verify this. There used to be a ford between Great Salkeld and Little Salkeld at NY 560367 (See map opposite page 171 in *CW2* xiii) and the field called Lady Fitts was possibly somewhere near this The word fitts or fit, common in neighbouring Edenhall Parish, is used of "grassland on the bank of a river" (ed. A. H. Smith, *English Place-Name Elements*, (Cambridge, 1970), part i, page 174).

page 97

1. Vulgare minus. **S** — *Lithospermum officinale* L.
2. Majus Dodonaei, flore purpureo.
3. Arvense radice rubente. Vide Anchusa degener.

Lolium. **Darnel.** Gall*ice* Ivray.
1. Album.
2. Rubrum.

Lonchitis vulgaris, Aspera. **Rough Spleen-wort. S** — *Blechnum spicant* (L.) Roth

Lunaria minor. **Small Moon-Wort.** T.L. et Fasc. At FARMANBY. HELS-FELL-wood. Near SETTLE, on the mountains. J.R. Fields 'twixt GLASSENBY and GAMELSBY. — *Botrychium lunaria* (L.) Sw.
Farmanby[7] NY 592371
Helsfell Nab SD 503936
Glassonby NY 578389
Gamblesby NY 610395

page 98

Lupulus. mas et foemina. **Hops. S** — *Humulus lupulus* L.

Luteola vulgaris. **Yellow-weld**; **wild-woad**, or **Dyers-weed.** Holm below LYNSTOCK-CASTLE. WETHERAL-ABBEY. etc. — *Reseda luteola* L.
Linstock Castle[8] NY 429585
Wetheral Abbey NY 468542

Lychnis. **Campion.** AS. Candel-wyrt.
1. Major Noctiflora
2. Flore albo minimo.
3. Alpina flore albo, Alsines myosotis facie.
3. Marina Anglica. Isle of FOWLEY. DUNNERHOLM nigh MARSH-GRAINGE,[9] FOURNESS. — *Silene maritima* With.
Foulney Island SD 247640
Dunnerholme SD 212798

---

7. This was the home of Thomas Lawson's daughter, Hannah (1668–1713), who married Isaac Thompson of Farmanby whose Quaker convictions contributed to his ruin in 1694 (R. Denton Thompson, "Farmanby and the Thompson family", *CW2* lv 179–190). Hannah is the "Mrs Thompson" of Thomas Robinson's plant list (See above pages xli–xliii).
8. This was the original residence of the Bishops of Carlisle. It was abandoned as such in 1450, but remained in ecclesiastical possession until 1863 (J. F. Curwen, *The Castles and Fortified Towers of Cumberland, Westmorland, and Lancashire North-of-the-Sands*, (Kendal, 1913), pages 298–299) The tithes of Linstock were leased to Nicolson's grandmother Radigunda Scott of Park Broom and accrued to the Nicolson family for several generations (*CW2* i 48). At the time of Nicolson's death, the lessee was his daughter, Catherine (*CW1* iv 10). For a note on "holm", see above n 11 page 10).
9. See above n 14 page 11.

4. Plumaria. Vide Armerius pratensis.
6. Viscosa ⟨1.⟩ alba Latifolia.
⟨2.⟩ rubra Angustifolia.
⟨3.⟩ flore muscoso.
5. Sylvestris.
1. flore Albo. **S** — *Silene alba* (Mill.) E.H.L. Krause
2. FLORE RUBRO. **S** — *Silene dioica* (L.) Clairv.

Vide Saponaria. et Behen Album. Nigellastrum.

Lysimachia. **Willow-herb**; or **Loose-strife.** page 99

1. galericulata. MAJOR et Minor. [T.L.] On the Mill-dam at CARLTON.[11] ISAN PARLES[12] HORNBY-Holm.[13] — *Scutellaria galericulata* L.[10] Isan Parles Caves NY 562303
2. Lutea vulgaris. Nigh BOWNAS, by WINANDERMEER. Willy-Holm at CARLILE. — *Lysimachia vulgaris* L. Willow Holme NY 393561
3. Lutea, flore globoso. In the East-rideing of YORKSHIRE; found by Mr DODESWORTH. J.R. — *Naumburgia thyrsiflora* (L.) Rchb.
4. purpurea spicata. Near the Castle-Mill at BROUGHAM. **S** In the marshy part of the HOLM.[14] — *Lythrum salicaria* L. Castle Mill NY 537292
5. Purpurea trifolia, caule hexagono. [Seu folio tripartite]. **S** In the fields about BLENNERHASSET, ASPATRICK, etc. Copiose. — *Lythrum salicaria* L. Blennerhasset NY 178415 Aspatria NY 147420
6. Siliquosa glabra. **Major** et **minor. Angustifolia** et †**Latifolia.**[15] **S**

---

10. There are two plants here, *Scutellaria galericulata* and *S. minor*, but Nicolson's underlining suggests that he wished only to mark down the former.
11. There is a Corn and Snuff Mill on the R. Eamont marked on the O.S. six-inch map of 1867 at NY 525288, quarter of a mile south of Carleton Hall.
12. For an account of these caves see A. J. Heelis, "The Caves Known as 'Isis Parlis'" *CW2* xiv 337–42.
13. Not located, unless this is one of the two small islands in the R. Eamont shown on the Tithe map of Brougham Parish, 1839, below Hornby Hall at NY 570300. See also n 11 page 10.
14. See above n 11 page 10.
15. It is difficult to know what Nicolson had in mind here. The plant that Ray found on the Cheviots and which he calls "Lysimachia siliquosa glabra minor latifolia" was probably *Epilobium alsinifolium* (Raven, *Ray*, page 155) If one were to assign modern names to the others, they would be *E. montanum* (Lysimachia siliquosa glabra major latifolia), *E. tetragonum* (—— —— —— major angustifolia) and *E. palustre* (—— —— —— minor angustifolia). The condensed reference, however, suggests that these were by no means so clearly separated in Nicolson's mind.

	†CHEVIOT-Hills, J.R. Also on HARTSIDE, CROSSFELL, etc.	Hartside Pass NY 647419 Cross Fell summit NY 687343
	7. Hirsuta. flore **Majore**, et **minore. S**	*Epilobium hirsutum* L. *Epilobium parviflorum* Schreb.
	8. Speciosa. Quibusdam Onagra siliquosa. **Rose-bay-Willow-Herb.** Long SLEDDALE. AUSTON-moor. Kirk-Haugh. Randal-Holm. Knaresdale, etc.	*Chamaenerion angustifolium* (L.) Scop. Alston Moor Kirkhaugh NY 698497 Randalholm Hall NY 708486
page 100	**M**	
	Malva. **Mallow** Vide ALTHAEA.	
	1. Vulgaris, procerior. **S**	*Malva sylvestris* L.
	2. Montana. Park.[16] NB. In the Lane 'twixt CARLILE and CARLTON.	Carleton NY 428528
	2. Sylvestris pumila. folio albo. **S** In the town-street. BRIGSTEER, town, etc, inter Rudera.	*Malva neglecta* Wallr. Brigsteer SD 482896
	3. Arborea marina nostras. BASSE-Island. J.R.	*Lavatera arborea* L. Bass Rock NT 603874
	4. Verbenacea. Vide Alcea.	
	Malus Sylvestris. Major et Minor. Albus et Ruber. **Crab-tree.**	
	Marrubium. **Horehound.** Vide Stachys. Sideritis.	
	1. Album, vulgare. **S**	*Marrubium vulgare* L.
	2. Nigrum. Vide Ballote.	
	3. Aquaticum. [T.L.]. HAWKSHEAD-moss,[17] At the Mill at BROUGHAM-Castle.	*Lycopus europaeus* L. Castle Mill NY 537292
page 101	Matricaria. Vide Parthenium.	
	Matrisylva. Vide Periclymenum et Asperula.	
	Medica. Vide Trifolium Luteum Syl*vestre*.	

16. A description of this plant is given on page 299 of John Parkinson's *Theatrum Botanicum*, (1640). It is possibly a form of *Malva moschata*.

17. Moss or bog land, now drained, lies east of Hawkshead although I have not found it named as Hawkshead Moss on any map.

Melampyrum. **Cow-wheat.**

1. Vulgare Sylvaticum, flore luteo. **S** In the Copse on TORPEN,[18] plentifully. — *Melampyrum pratense* L. Torpenhow village NY 203397
2. Cristatum, Angustifolium.

Melilotus, sive Trifolium odoratum. **Melilot.**

1. Vulgaris, flore Luteo. **S** BY THE RIVER-SIDE. [...] ON THE BANK UNDER LANGWATHBY, copiose. — *Melilotus altissima* Thuill. Langwathby NY 570335
2. Coronaria, vel Coronata.[19] Vide Trifolium corniculatum. **S**

Mentha. **Mint.**

page 102

1. Aquatica. folio –
   1. Glabro.[20] **S**
   2. Hirsuto.
2. Sylvestris; sive Menthastrum. **Horse-mint.** HELSTON-LATHES;[21] nigh the Kiln. FLOUKBARROW near CARTMEL. Sylv*aticum* Spicatum. — *Mentha rotundifolia* (L.) Huds. Helsington Laithes SD 505906 Flookburgh SD 367759
3. Cattaria; sive Nepeta vulgaris. **Nep**, or **Catmint.** In the Abbey at CARLILE,[22] and about the walls. — *Nepeta cataria* L.
4. Angustifolia Spicata; seu Romana. **Spear-mint**; or **Heart-mint.** Injucundior Mentha Cardiaca hortorum.

Mercurialis, Mas et Foemina. **French Mercury.**

Montana. Vide Cynocrambe.

Meum vulgare; sive Radix Ursina. **Common Spignel, or Meu.** [T.L]. In a Corn-Close — *Meum athamanticum* Jacq.

page 103

---

18. As a good Saxonist, Nicolson drops the superfluous "how" from the end of the name. *Nicolson and Burn*, vol ii, page 124, have an interesting note on the tautological formation of the name. Nicolson, who is referring here to the hill above the village, did in fact hold the living of Torpenhow from 1681 to 1698, although he never lived there. In 1698, he resigned it in favour of his brother-in-law, Thomas Nevinson.
19. In general terms, this is Birdsfoot-trefoil. Under "Trifolium Corniculatum", to which he gives a cross-reference, Nicolson was to particularise these.
20. Impossible to identify with any confidence, though possibly *Mentha aquatica* or *Mentha* × *verticillata.*
21. See above n 86 page 95.
22. See above n 37 page 66.

near KENDAL-Castle. About SHAP. Nostras BAUDMONEY. — Kendal Castle SD 522924

Militaris Aizoides. **Water Sen-green**, or **Fresh-water-soldier.**

Milium Solis. Vide Lithospermum.

Millefolium. **Milfoil**; or **Yarrow.**

1. Album, vulgare; Terrestre. **S** — *Achillea millefolium* L.
2. Aquaticum; sive viola Aquatica. BRIGSTEER-MOSS. — *Hottonia palustris* L. Brigsteer village SD 482896
3. Aquaticum cornutum.
4. Pennatum aquaticum. KENDALL: highest Tarn above the Tenters.[24] — *Myriophyllum spicatum* L.[23]
5. palustre galericulatum majus et minus. 'Twixt CROOK and WINANDERMEER. Nigh WITHERSLACK. — *Utricularia vulgaris* L. and *U. minor* L. Crook SD 461950 Witherslack SD 431841
6. Ranunculi flore. **Water Fennel. S** — *Ranunculus aquatilis agg.*
7. Terrest⟨r⟩e flore rubro. qu. an diversum a *primo*.

page 104 Millegrana Minima; sive Herniaria minor. **Least Rupture-wort**; or **All-seed. S** Upon most moors in the places where the outer Turff has bin shaved off.[25] — *Radiola linoides* Roth

Mollugo. Vide Gallium et Rubia.
Montana. Major et Minor. **Bastard-Madder. S** — *Galium boreale* L. and *G. saxatile* L.

Morsus Diaboli; sive succisa. flore Scab*iosae* purpureo. **Devil's-bit. S** — *Succisa pratensis* Moench

Morsus Gallinae. Vide Alsine [Hederacea] \vulgaris./et Lamium.

Morsus Ranae. Vide Nymphaea.

Muscipula Salamantica. Vide Sesamoides.

23. Probably includes other species.
24. See above n 29 page 16.
25. As a preliminary to the cutting of peat.

Entry	Identification	MS page
Muscus. **Moss.** Graece Μόσχος[26], Novellum et Tenerum.		page 105
1. Aquaticus terrestri similis.		
1. Arboreus. **Tree-moss.**		
1. Ramosus, vulgaris. **S**	*Usnea* sp., possibly *plicatus*	
2. Peltatus et Scutellaris; floridus.		
3. Nodosus; seu geniculatus. J.R.[27] at BURNLEY in LANCASHIRE Near KENDAL. T.L.	*Usnea* sp., possibly *barbata*	
4. Supinus marginibus pilosis.		
5. Villosus; Usnea officinarum.		
2. Crustae (aut Lichenis) modo Arboribus adnescens. Qui vel –		
1. Flavus.		
2. Cinereus.		
3. Pullus. Fasc.		
3. Capillaris. Vide Adianthum Aureum.		
4. Clavatus; sive Lycopodium. **Club-moss**; or **Wolfes-claw. S** Plicaria, POLONIS.		
1. Lycopodium elatius juniperinum.	*Lycopodium clavatum* L.	
2. folijs retro reflexis.		page 106
3. Terrestris Repens. T.L. SHAP-moor.	*Lycopodium inundatum* L.	
4. Folijs Cupressi. T.L. BUCKBARROW-well.	*Lycopodium alpinum* L. Buckbarrow Well NY 478076	
5. Erectus Abietiformis. T.L. **S** [. . .]	*Lycopodium selago* L.	
6. Corniculatus. **S** Q. tamen an TABERN.	*Cladonia* sp.	
7. Coralloides apicibus coccineis. **S**	*Cladonia pyxidata*	
8. Croceus saxatilis, serico similis.		
8. Denticulatus minor.		
9. Filicinus. **Fern-moss.**		
10. Lanuginosus. vitium graminis.		
10. Marinus.		
1. Albus, vulgatissimus. DUNNERHOLM in FOURNESS.	Dunnerholme SD 212798	
2. Coralloides denticulatus.		

26. The young shoot of a plant, a sprout, sucker. The Latin is really a gloss of the Greek.
27. "T.L." has been altered here to "J.R."

3. Corallinae in modum articulatus. Fasc.
4. Denticulatus minor.

page 107

5. Denticulatus procumbens, caule tenuissimo.
6. Denticulis bijugis unum latus spectantibus.
7. Equisetiformis non Ramosus. **Mrs. Ward's Sea-beard.** GISBURGH. — Guisborough
8. Erectior, Ramulis in innumera et tenuissima Capillamenta divisis.
9. Pennatus, ramulis falcatis. SICKLE-FEATHER'D.
9. purpureus parvus.
10. Rubens pennatus. BERWICK. DUNNERHOLM in FOURNESS. — Dunnerholme SD 212798
11. Palustris; Terrestri similis.
12. Polyspermos. **Seeding mountain-moss. S** BUCKBARROW-well, etc. T.L. — *Selaginella selaginoides* (L.) Link; Buckbarrow Well NY 478076
13. Pyxioides. **Cup-** or **Chalice-moss. S** Item pyxidatus multiformiter apicibus coccineis. D*octor* Plott. — *Cladonia pyxidata*
14. Parvus stellaris. **Small Heath-moss.**
15. Terrestris
    1. Major.
    2. Minor; Adianthi aurei Capitulis.

page 108

16. Tubulosus Ramosissimus fruticuli specie; sive Muscus Corallinus Montanus. Albus.
17. Trichomanoides purpureus, Alpinis rivulis innascens. Near SKIPTON J.R.

Myagrum Sylvestre. **Wild Gold of pleasure.** L.[28] — *Camelina sativa* (L.) Crantz
Siliqua longa. Treacle-wormseed.

Myosotis Scorpoides.
Mouse-Ear-Scorpion-grass.
1. Arvensis Hirsuta. **S** — *Myosotis arvensis* (L.) Hill
2. Minor, flosculis luteis.
3. Palustris. **S** — *Myosotis scorpioides* L.

---

28. See above n 27 page 15.

Myosuros. **Mouse-tail.** CARTMEL-Grainge.[29] *Myosurus minimus* L.

Myrrhis Sylvestris. Vide Cicutariae [vulgaris]\simile/Cerefolium.

Myrtus Brabantica. Vide Elaeagnus Cordi.

N

page 109

Napus Sylvestris. **Wild Navew.** forte, **Rape**. J.R.

Narcissus Sylvestris, Luteus. **Wild Daffodil**, or **Daffadowndilly.** In the grounds about Dockraw-Hall[30] near KENDALL. *Narcissus pseudonarcissus* L.

Medio-luteus. **Primrose Peerless**; or **common pale daffodil.** cum priore *non* raro. *Narcissus* × *biflorus* Curt.

Nasturtium.

1. Aquaticum amarum. **Bitter cresses.** T.L. in watery places in SELSIDE. Baggra-lane[31] near ULNDALE. *Cardamine amara* L. Selside village SD 527993
2. Aquaticum, folijs minoribus precocius. **S** *Rorippa nasturtium-aquaticum* (L.) Hayek.
3. Petraeum. By Common-holm-bridge, Great STRICKLAND. *Teesdalia nudicaulis* (L.) R.Br. Commonholme Bridge NY 576247
3. Pratense. Vide Cardamine.
4. Hybernum. Vide Barbarea.
5. Sylvestre Erucae affine.
6. Verrucosum. Vide Coronopus Ruel*lij*.

Nepeta vulgaris. Vide Mentha Cattaria.

page 110

Nidus avis. Vide Satyrion abortivum.

Nigellastrum. Vide Pseudomelanthium.

---

29. See above n 3 page 6.
30. This would be Dockwra Hall, which originally took its name from the family owning it, of whom Cornelius Nicholson gives an account in *Annals of Kendal*, 2nd ed. (London and Kendal, 1861), pages 79–80. The position was SD 513934.
31. Probably the lane serving Baggra Yeat Farm, NY 267367.

Nucula Terrestris. Vide Bulbocastanum.

Nummularia. **Money-wort**; or **Herb Two-pence**.

1. Major vulgaris, lutea. BRIGSTEER. — *Lysimachia nummularia* L. Brigsteer SD 482896
2. Minor; flore purpurascante T.L. In the marshes the foot-way 'twixt CROOK and WINANDER-MEER. **S** — *Anagallis tenella* (L.) L. Crook SD 461950

Nux vesicaria. Vide Staphylodendron.

Nymphaea. **Water-Lily**.

1. Alba
   1. Major, vulgaris. In stagnis, about STANWIX, etc. — *Nymphaea alba* L. Stanwix NY 400570
   2. Minor, seu Morsus Ranae CLIBBURN. — *Hydrocharis morsus-ranae* L. Cliburn NY 588245
2. Lutea. Major etiam et Minor. cum priore. HAWKSHEAD-Moss;[32] in the River. — *Nuphar lutea* (L.) Sm.

page 111 **O**

Oculus Christi. Vide Horminum Sylvestre.

Ocymum Cereale. Vide Fegopyrum.

Oenanthe. [Vulgaris Vide Filipendula]. DROPWORT.

1. Vulgaris. Vide Filipendula.
2. Aquatica
   1. Major. 'Twixt MARSH-GRANGE[33] and the I*sle* of WALNEY. — *Oenanthe fistulosa* L. Marsh Grange SD 220797
   2. Minor.
3. Cicutae facie. **S** Nostrat*ibus* DEAD-TONGUE. — *Oenanthe crocata* L.

Onobrychis. **Medick-vetchling**; or **Cocks-head**. Sainct foin. At Dunnerholm in FOURNESS. — *Onobrychis viciifolia* Scop. Dunnerholme SD 212798

---

32. See above n 17 page 58.
33. See above n 14 page 11.

Entry	Modern identification	Page
Ononis. Vide Anonis.		
Onopyxos. Vide Carduus Asininus.		
Ophioglossum. **Adders tongue**. T.L. HELSFELL-woods. **S**	*Ophioglossum vulgatum* L. Helsfell Nab SD 503936	page 112
Ophrys. Vide Bifolium.		
Orchis.		
1. Abortiva russa, sive Nidus avis. **Birds-nest**. In a wood 'twixt CONSWICK and STRICKLAND-Roger. In a Bog in Sir J.L.\'s/wood a *quarte*r of a mile below STRICKLAND-Mill.	*Neottia nidus-avis* (L.) Rich. Cunswick Scar SD 491940 Strickland Mill NY 551225	
2. Abortiva violacea. **Purple Birds-nest**.		
3. Arachnitis. **Spider-Orchis**.		
4. Barbata foetida. **Lizard-flower** or **Great Goatstones**.		
5. Galea et alis fere Cinereis. **Man-orchis**.		
6. Hominis nudi effigiem flore repr⟨ae⟩sentans. Fasc.		page 113
7. Lilifolius minor Sabuletorum Zelandiae et Bataviae. **S**	*Liparis loeselii* (L.) Rich.	
8. Morio mas, folijs maculatis. Male **Foolstones**; or **Crow-toes**. **S**	*Orchis mascula* (L.) L.	
9. Morio foemina.		
10. Myodes. **Fly-orchis**. T.L. In Sir Ch. Philipson's grounds at HELSFELL. STRICKLAND-Rogers,[34] nigh KENDAL.	*Ophrys insectifera* L. Helsfell Nab SD 503936	
11. Militaris Pannonica. **Little purple flower'd Orchis**; or **Lame-Soldier-Orchis**. T.L. **S** In the Holm by the Mill.[35]	*Orchis ustulata* L.	
12. Palmata. **Handed Orchis**.		
1. Major, Mas. **Satyrion Royal, male**.		
2. Foemina. **Female Satyrion Royal**.		
3. minor, flore rubro. **S**	*Gymnadenia conopsea* (L.) R.Br.	
4. flore viridi.		

34. The manorial lands of Strickland Roger lay some 2 miles north-easterly from Cunswick Scar.
35. The mill at Great Salkeld, called Force Mill, was in a position NY 561379. *CW2* xiii 170–171. For holm land, see n 11 page 10.

page 114

13. Purpurea spica congesta pyramidali.
14. Pusilla odorata, Lutea. **Musk-Orchis**.
15. Serapias bifolia vel trifolia minor. **Butterfly-Satyrion**.
16. Sphegodes; sive fucum referens. **Humble-Bee-Satyrion**.
17. Spiralis alba, odorata. **Triple Ladies-traces. S** — *Spiranthes spiralis* (L.) Chevall.

Origanum vulgare, spontaneum. **Wild Marjoram. S** — *Origanum vulgare* L.

Ornithogalum. **Star of Bethlehem**.
1. Angustifolium majus; floribus ex albo virescentibus.

page 115

2. Luteum; sive Caepe Agraria. **S** WILLIES-WOOD,[36] plentifully. GREATA-BRIDGE. — *Gagea lutea* (L.) Ker-Gawl. Greta Bridge NZ 086132
3. vulgare.

Ornithopodium. **Birdsfoot**; or **Birds-claw**. About the Tenters[37] at KENDALL. **S** in the BARWGH,[38] etc. — *Ornithopus perpusillus* L.

Orobanche, flore majore. **Broom-Rape**. T.L. DODDING-green, nigh KENDALL. Copiose. CARLTON-Lane, nigh CARLILE; etc. Vide Orchis abortiva. Dentaria. — *Orobanche rapum-genistae* Thuill. Dodding Green SD 534954 Carleton NY 428528

Orobus Sylvaticus Nostras. **English Orobus**. T.L. et J.R. **S** Copiose. But especially at BLENCARN. In SCOTLAND. Sutherl. Nostras HORSEPEASE [Vide Orchis abortiva. Dentaria.] — *Vicia orobus* DC. Blencarn NY 636313

Osmunda Regalis. Vide Filix florida.

Ostrys. Vide Betulus.

Oxyacantha. **Barberries**; or **Pipperidges**.

---

36. See above n 44 page 23.
37. See above n 29 page 16.
38. Not located, but probably in Great Salkeld Parish. John Ray, *A Collection of English Words Not Generally used*, (1674), gives "A Bargh, A Horseway up a steep hill. York-shire."

Oxycanthus. **White-thorn**; or **Haw-thorn**. **S** — *Crataegus monogyna* Jacq.

Oxalis. Vide Acetosa.

Oxymyrsine. Vide Ruscus.

Oxys Plinij. Vide Acetosum trifolium.

P

page 116

Padus. i.e. Cerasus avium.

Palma Christi. Vide Orchis palmata.

Paludapium. Vide Apium palustre.

Panax coloni. Vide Sideritis Anglica.

Papaver. **Poppey**. Vide Argemone.

1. Corniculatum luteum. On the Sea-shore, frequens. Isle of FOWLEY. — *Glaucium flavum* Crantz Foulney Island SD 247640
2. Corniculatum violaceum.
3. Erraticum Campestre Rubrum. 'ῥοιὰς[39] Diosc. Near PENRITH-Castle. **S** — *Papaver rhoeas* L. Penrith Castle NY 513298
3. Capitulo longiore glabro.
4. Spontaneum Sylvestre.
5. Spumeum. Vide Behen Album.

Paralysis. Vide Primula Veris.

page 117.

Parietaria. **Pellitory of the Wall**. Abundantly on the walls at CARLILE. — *Parietaria diffusa* Mert. & Koch

Paronychia. **Chickweed**; or **Whitlow-grass**.

1. Vulgaris. **S** — *Erophila verna* (L.) Chevall.
2. Folio Rutaceo. On old walls at NEWBY, SHAP. etc. — *Saxifraga tridactylites* L. Newby NY 590213

Parthenium; sive Matricaria vulgaris. **Feverfew**. AS. Adre-mint. Under the walls at CARLILE, etc. — *Chrysanthemum parthenium* (L.) Bernh.

Pastinaca. **Parsnep**.

1. Latifolia sativa. **Common Parsnep**. **S** — *Pastinaca sativa* L.
2. Sylvestris. Vide Elaphoboscum et Daucus.

---

39. wild poppy.

page 118	2. Tenuifolia Sativa. **Carrots**.	
	1. Radice Lutea et Alba. **S**	*Daucus carota* L.
	2. Radice Atro-Rubente. **S**	*Daucus carota* L.
	3. Aquatica. Vide Sium.	
	Pecten Veneris; seu Scandix vulgaris. **Shepherds needle**; or **Venus-comb**. About Great STRICKLAND. Near St NICHOLAS, CARLILE.[40]	*Scandix pecten-veneris* L. Great Strickland NY 560230
	Pedicularis, **Lowsewort**; or **Rattle**.	
	1. Lutea; seu Crista Galli. **Yellow Rattle**; or **Cockscomb**. **S**	*Rhinanthus minor* L.
	2. Pratensis Rubra vulgaris. **S**	*Pedicularis sylvatica* L.
	3. Palustris Rubra, elatior. **S**	*Pedicularis palustris* L.
page 119	Pentaphyllum. **Cinquefoil**. Vide Tormentilla.	
	1. Erectum, folio Argenteo. **Tormentil**, or **Wall-Cinquefoil**. Long-SLEDDALE.	*Potentilla argentea* L.
	2. Incanum Repens Alpinum. About BUCK-BARROW.	*Potentilla tabernaemontani* Aschers. Buckbarrow Crag NY 482075
	3. Rubrum palustre. **S**	*Potentilla palustris* (L.) Scop.
	4. Vulgatissimum, repens, flore luteo. **S**	*Potentilla reptans* L.
	Pentaphylloides fructicosum. **Shrub-Cinquefoil**. Qu. An distinctum a pentaphyllo fragifero Fasciculi? On the banks of TEES; about EGLESTON-ABBEY, etc. In an Isle in the same River over against MICKLETON.	*Potentilla fruticosa* L. Eggleston Abbey NZ 062150 Mickleton NY 967237
	Peplus; seu Esula rotunda. **Petty Spurge**. maritima folio obtuso.	
page 120	Percepier Anglorum. Sive, Alchimilla montana minima, **Parseley-piert**. T.L. **S**	*Aphanes arvensis* L.
	Perfoliata. **Thorow-wax**.	

---

40. The Hospital of St. Nicholas, which was founded before the reign of King John, was destroyed during the seige of Carlisle in 1645. See *V.C.H.*, Cumberland, (1905), vol ii, pages 199–203. The area, a mile south-easterly of the city centre and now built over, is still known as St. Nicholas.

Perforata. Vide Hypericum.

Periclymenum. **Woodbind**; or **Honey-suckles**.
1. Non perfoliatum. Seu, Caprifolium vulgare. **S** — *Lonicera periclymenum* L.
2. Parvum pruterium. Vide Chamaepericlymenum.

Persicaria. **Arsmart**.
1. Maculosa, mitis. **S** — *Polygonum persicaria* L. — page 121
2. Acris, urens; seu Hydropiper. **S** — *Polygonum hydropiper* L.
3. Pusilla repens.
4. Siliquosa; sive, Noli me tangere. **Codded Arsmart**. T.L. 'Twixt AMBLESIDE and RYDALE. HAWKESHEAD. — *Impatiens noli-tangere* L.
5. Salicis folio perennis. i.e. Potamogiton Angustifolium.

Pes Anserinus. Vide Atriplex Sylvestris.

Pes Cati. Vide Gnaphalium montanum album.

Pes Columbinus, Vide Geranium columbinum.

Petasites vulgaris; sive Tussilago major. **Butter-bur**; or **pestilent-wort**. **S** — *Petasites hybridus* (L.) Gaertn., Mey. & Scherb. — page 122

Peucedanum. **Hogs fennel**; **Sulphurwort**; or **Harestrong**.
1. Vulgare.
2. Minus. **Rock-persly**.

Phalaris pratensis. Vide Gramen tremulum.

Phellandrium. Vide Cicutaria palustris.

Phoenix. Vide Lolium rubrum.

Phu. Vide Valeriana.

Phyllitis; sive Lingua Cervina. **Hartstongue**. Gr*aece* Φυλλίτις,[41] i.e. Foliosa. — page 123

---

41. Leafy. The Latin glosses the Greek.

	1. Vulgaris. LEVENS-Mill	*Phyllitis scolopendrium* (L.) Newm.
	2. Lacinata. A priore specie *non* distincta. J.R. CARTMEL-well.[42] INGLEBOROGH	*Phyllitis scolopendrium* (L.) Newm.
	Pilosella. **Mouse-ear**.	
	1. Vulgaris, Repens. **S**	*Hieracium pilosella* L.
	2. Siliquata. Major et Minor.	
	Pimpinella. **Burnet**.	
	1. Sanguisorba. quae –	
	1. Vulgaris, Major. **S**	*Sanguisorba officinalis* L.
	2. Minor, Hortensis. **S**	*Poterium sanguisorba* L.
	Sanguisorba pilos, Saxifraga non habet ullos.	
page 124	2. Saxifraga. Quae etiam –	
	1. Major. **S**	*Pimpinella major* (L.) Huds.
	2. Minor. **S**	*Pimpinella saxifraga* L.
	3. Hircina media. Sive Daucus Selinoides Cordi.	
	Pinguicula, **Butterwort**; or **Yorkshire-Sanicle**.	
	1. Vulgaris, Gesneri. T.L. GREAT STRICKLAND. **S**	*Pinguicula vulgaris* L. Great Strickland NY 560230
	2. Flore minore carneo.	
	Pinus. **Pine-tree**.	
	Sylvestris folijs brevibus glaucis, conis parvis albentibus. **Scotch-Firr**. Fasc.	
page 125	Pisum. **Pease**.	
	1. Arvense. **Field-pease**.	
	1. Fructu albido.	
	2. Fr*uctu* Cinericeo.	
	3. fr*uctu* variegato. Maple-pease.	
	2. Maritimum Britannicum. Nigh WHITEHAVEN.	*Lathyrus japonicus* Willd.
	Plantago. **Plantaine**; or **Waybread**. AS. Wegbraede Vide Cornu Cervinum.	

42. See above n 13 page 11.

Entry	Identification	
1. Aquatica. quae vel –		
1. Major, Latifolia. **S**	*Alisma plantago-aquatica* L.	
2. Minor Angustifolia. BRIGSTEER. In the Sike near the Cloven stone[43] on great STRICKLAND-moor.	*Baldellia ranunculoides* (L.) Parl. Brigsteer SD 482896	
3. Minor Stellata.		
4. Minima. On STRICKLAND Common	Strickland Common Not located	
2. Vulgaris. Atque haec –		
1. Folio glabro, major. **S**	*Plantago major* L.	
2. Hirsuta, Latifolia. **Lambs-tongue**. **S**	*Plantago media* L.	page 126
3. Spiralis, pannicula Sparsa. **Besom-plantain**. Johns.		
4. Quinquenervia, Angustifolia. **Ribwort**. **S**	*Plantago lanceolata* L.	
3. Alpina Angustifolia.		
4. Marina. On *our* Shore; below Nether-LEVENS, etc. In the high way at APPLEBY town-end.	*Plantago maritima* L. Nether Levens SD 488851	
Pneumonanthe. Vide Gentianella palustris Angustifolia.		
Podagraria. Vide Herba Gerardi.		
Polyacantha. Vide Carduus Spinosiss*imus*.		
Polyanthemum palustre. Vide Ranunculus Aq*uatilis*.		
Polygala, Milkwort. **S** flore rubro, Albo et coeruleo.	*Polygala vulgaris* L.	
Polygonatum. **Solomon's seal**. On INGLEBOROUGH.	*Polygonatum multiflorum* (L.) All.	
flore majore odoro. On the Scars about SETTLE. J.R.	*Polygonatum odoratum* (Mill.) Druce	

43. Great Strickland Moor lies half way between Great Strickland and Morland. There are several sikes (ditches) in the area. Cloven Stone appears to exist no longer.

page 127	Polygonum. **Knot-grass**.	
	1. Germanicum tenuifolium; seu Knawel. **S**	*Scleranthus annuus* L.
	2. Germ*anicum* Incanum flore majore.	
	3. Marinum, +Latifolium. Et, minus folijs Serpylli. +On the shore 'twixt WORKINTON and WHITEHAVEN.	*Polygonum raii* Bab.
	4. Mas, vulgare.	
	5. Parvum. flore albo verticillato.	
	6. Selinoides. i.e. Percepier Anglorum.	
	Polypodium. **Polypody**.	
	1. Vulgare. **S**	*Polypodium vulgare* agg.
	2. Cambro-Britannicum, pinnulis ad marginem laciniatis. Fasc.	
page 128	Populus. **The Poplar Tree**.	
	1. Alba. vel folijs –	
	1. Majoribus, vulgaris.	
	2. Minoribus. **Abele-tree**.	
	2. Nigra. RENWICK-wood.	*Populus nigra* L. Renwick village NY 595435
	3. Tremula; seu Lybica. The **Asp-Tree**. **S** AS. Aeps. Aespe.	*Populus tremula* L.
	Portulaca. **Purslane**.	
	1. Aquatica. Vide Alsine rotundif*olia*.	
	2. Marina. Vide Halimus	
	Potamog[e]iton. **Pondweed**.	
page 129	1. Angustifolium. **S**	*Polygonum amphibium* L.
	2. folijs longis acutis, splendescentibus.	
	3. Folijs crispis. Seu, Lactuca Ranarum. Vide Tribulus.	
	3. Latifolium, vulgare. **S**	*Potamogeton natans* L.
	4. Latifolium Splendescens, perfoliatum.	
	5. Maritimum pusillum.	
	6. Pusillum gramineum.	
	7. Ramosum caule compresso.	
	8. Ramosum Gramineum.	
	Potentilla. Vide Argentina.	
	Prassium. Vide Marrubium album.	

Primula veris; sive Paralysis.		page 130
1. Major. **Cowslips**; or **Paigles**. **S**		
1. Odorata flore luteo. **S**	*Primula veris* L.	
2. Inodora. **Oxlips**. In the grassings[44] at HELSFELL. Nostras LAD-CANDLESTICKS. At the water-meetings 'twixt Caldbeck and Hesket.	*Primula veris* L. x *vulgaris* Huds. Helsfell Nab SD 503936 Junction of Cald Beck and R. Caldew NY 343399	
2. Minor.		
1. Vulgaris. **Common Primrose. S**	*Primula vulgaris* Huds.	
2. Flore Rubro. **Birds-een**. **S**	*Primula farinosa* L.	
Prunella. **Self-heal S**[45]	*Prunella vulgaris* L.	
Prunus Sylvestris.		
1. Fructu majori. **Bullace-Tree**. Hocque		
1. Albo. About AMBLESIDE.	*Prunus domestica* L. *ssp insititia* (L.) C.K. Schneid.	
		page 131
2. Nigro.		
3. Rubro.		
2. Fructu minore. **The Black-thorn**; or **Sloe-Tree**. Prunellum, officin*arum*. **S**	*Prunus spinosa* L.	
Pseudomelanthium. **Cockle**. In the corn about LINSTOCK, CARLILE, etc. **S** qu. an Lychnis Segetum parva viscosa NEWTONI, ab hoc differat specie?	*Agrostemma githago* L. Linstock Castle[46] NY 429585	
Pseudo-Narcissus. Vide Narcissus.		
Ptarmica. **Sneezewort**; or **Bastard pellitory**. Nostras Goose-tongue.		
1. Vulgaris. **S**	*Achillea ptarmica* LV.	
2. Flore pleno. T.L. WINANDERMEER, in the small holm.[47]	*Achillea ptarmica* L.	

---

44. See above n 44 page 23.
45. Three words in German script follow, possibly "von der Bräune". See *O.E.D.* entry for *Prunella*.
46. See above n 8 page 56.
47. Nicolson clearly has a specific island in mind (see also above page 9) but the name Small Holme has not survived.

	Pulegium. **Penny-Royal**; or **Pudding-grass**. In a watery place in RAMPSIDE in FOURNESS.	*Mentha pulegium* L. Rampside SD 240663
page 132	Pulmonaria. **Lungwort.**	
	1. vulgaris [...] Vide Lichen Arborum.	
	2. folijs Echij, flore rubro. **Bugloss-Cowslips**; or **long-leav'd Sage of Jerusalem**. Nigh BURNISHEAD-Chapple. J.F.R.	*Pulmonaria longifolia* (Bast.) Bor. Burneside Chapel[48] SD 504957
	3. Gallica; sive Aurea.	
	1. Latifolia. **S**	
	2. Angustifolia.	
	Pulsatilla [vulgaris]\Anglica/Purpurea.	
	Pyrola. **Winter-green**.	
	1. Alsines flore Europaea. On RUMBLESMEER[49] near HELWICK. On the Picts-wall near HEXHAM.	*Trientalis europaea* L.
page 133	2. Vulgaris Rotundifolia nostras. About HALIFAX and KEIGHLEY.	prob. *Pyrola media* Sw.
	3. Alsines flore Brasiliana. T.L. in Fasc. near GISBURGH in CLEVELAND.	Guisborough
	4. folio mucronato serrato. HASELWOOD in YORKSHIRE.	*Pyrola minor* L. Hazlewood Hall SE 089539
	Pyrus Sylvestris. **Wild Pear-tree**.	

Q

Quercus. **Oak**. AS. Aac, Ac. Danis, eik.	
1. Vulgaris, Latifolia. **S**	*Quercus robur* L.
2. Marina. **Sea-oak**; or **Wrack**. Vide Fucus et Alga.	
Quinquefolium. Vide Pentaphyllon.	

48. *PNW* i 153.

49. Rumblesmeer is Rombalds Moor, the high ground between the Skipton–Otley road and the Skipton–Shipley road, in the West Riding. Helwick is Eldwick (*PNY:W* iv 162) at SE 125403.

R

page 134

Radix cava minima flore viridi. **Tuberous Moscatell. S**	*Adoxa moschatellina* L.
Ranunculus. **Crow-foot.** Vide Millefolium Aqu*aticum*.	
1. Aquatilis.	
2. Arvensis. In the corn about MORLAND.	*Ranunculus arvensis* L. Morland NY 600225
3. Auricomus.	
4. Bulbosus. **Batchelour's Buttons**.	
5. Flammeus, major et **minor**.[50] Spear-wort. **S**	
6. Flammeus Serratus. **S**	*Ranunculus flammula* L.
7. Globosus. **Locker-gowlons. S**	*Trollius europaeus* L.
8. Hederaceus Rivulorum, atra macula notatus. About MARSH-GRAINGE in FOURNESS.	*Ranunculus hederaceus* L. Marsh Grange[51] SD 220797
9. Nemorosus Dulcis. **S** [i.e. Radix cava].	*Ranunculus auricomus* L.

page 135

10. Palustris Rotundifolius. folio Apij. In the Cittadel-pond at CARLILE.[52] Hell-kettles nigh BLACKWELL.	*Ranunculus sceleratus* L. Hell Kettles[53] NZ 281109
11. Pratensis Repens. **Butter-Cups**.	
12. Pratensis erectus acris.	
13. Rectus, Hirsutus.	
1. folijs pallidioribus.	
2. Annuus, flore minimo.	

---

50. There appear to be two plants here: *Ranunculus lingua* ("Ranunculus flammeus major") and *R. flammula* ("Ranunculus flammeus minor"). The fact that Nicolson has underlined only "minor" would appear to indicate that he reckons only to have seen the latter species, although his duplication of it in the next entry leaves one in some doubt as to what he actually saw.
51. See above n 14 page 11.
52. This has long since disappeared. The Citadel guarded the English gate at the south-east angle of the city walls, very near to the present Citadel Railway Station (NY 402556). It was largely demolished when, in 1807, an act was passed to enable the building of the Assize Courts on the site (Samuel Jefferson, *The History and Antiquities of Carlisle*, (Carlisle, 1838), pages 277–79).
53. R. Taylor Manson, *Zig-Zag Ramblings of a Naturalist*, (Darlington, 1898), pages 75–99 gives an account of Hell Kettles, together with some notes on the botany of the locality. These circular hollows, filled with water were traditionally supposed to be bottomless and to be connected with the Tees by an underground passage. They are mentioned by both Leland and Camden.

Page	Entry	Identification
	Raphanus. **Radish**.	
	1. Aquaticus. **Water-Radish**.	
	2. Marinus. i.e. Eruca marina.	
	3. Rusticanus; seu Sylvestris. **Horsh-Radish**. [*sic*] Mr Smalwood's Hill at *Little* SALKELD. Near ALNWICK, BOLLAND in CRAVEN, etc. J.R.	*Armoracia rusticana* Gaertn., Mey. & Scherb. Mr Smalwood's Hill not located Forest of Bowland[54]
page 136	Rapistrum. **Charlock**; or **Chadlock**.	
	1. Articulatum Album.	
	2. Arvorum, flore Luteo. **Wild Mustard. S**	*Sinapis arvensis* L.
	3. Siliqua glabra articulata.	
	Rapum. **A Turnep**.	
	1. Sativum rotundum.	
	2. Radice oblonga.	
	3. Sylvestre.	
	Rapum Genistae. Vide Orobanche.	
page 137	Rapunculu⟨s⟩ **Rampions**.	
	1. Corniculatus montanus, Spica orbiculari.	
	2. Esculentus vulgaris.	
	3. Scabiosae capitulo coeruleo. **S**	*Jasione montana* L.
	Regina Prati. Vide Ulmaria.	
	Reseda vulgaris. **Italian Rocket**.	
	Rhamnus Catharticus. **Buck-thorn**.	
	1. Vulgaris. T.L. GREAT STRICKLAND, and SHERIF'S-PARK. **S**	*Rhamnus catharticus* L. Great Strickland NY 560230 Sheriff's Park NY 554214
	2. Litoralis. **Sallow-thorn**; or **Sea-Buckthorn**. 'Twixt WHITBY and LYTH.	*Hippophaë rhamnoides* L.

54. *PNY:W* vi 112.

Rhodia Radix. **Rosewort; or Rose-root**. J.R. et T.L. HARDKNOT; and HARTERFELL-CRAG, near MARDALE. INGLEBOROUGH.	*Sedum rosea* (L.) Scop. Hardknott Pass NY 231014 Harter Fell summit NY 460094	page 138
Ribes. **Currans**.		
1. Vulgare, fructu rubro. In the woods of YORKSHIRE, WESTMERLAND and DURHAM. J.R.	*Ribes rubrum* agg.	
2. Fructu nigro. **Black Currans**; or **Squinancy-berries**. **S**	*Ribes nigrum* L.	
3. Alpinus dulcis.		
Rorella. Vide Ros solis.		
Ros solis, **Sun-dew**; **Rosa solis**, or **Redrot**.		
1. folio oblongo.		
2. folio rotundo. **S** In Mr Richardson's [Bogg] BIRCH-CLOSE.[55] Common in most Mosses. Nostras Moor-grass.	*Drosera rotundifolia* L.	
3. Rotundifolia perennis.		
4. Perennis longifolia. 'Twixt DONCASTER and BAUTREY.	*Drosera intermedia* Hayne.[56] Bawtry	
5. Longifolia maxima. Near CARLILE.	*Drosera anglica* Huds.	
Rosa. A **Rose**. Sylvestris species *sunt* quae sequuntur –		page 139
1. Pimpinellae folio. **S**	*Rosa pimpinellifolia* agg.	
2. Pomifera major. **S**	prob. *Rosa villosa* agg.	
3. Inodora, seu Canina. **Dog's Rose**; or **Briar-bush**. **S**	*Rosa canina* agg.	

55. The Richardsons were the owners of Nunwick Hall in the 17th century and were related to the Huttons of Penrith (Loftie, *Salkeld*, pages 90–94). The Richardson referred to is probably Christopher Richardson, who is noted by Nicolson in his terrier of 1703–04 (ed. R. S. Ferguson, *Miscellany Accounts of the Diocese of Carlile, with the Terriers delivered in to me at my Primary Visitation*, (London and Carlisle, 1877), page 184) as paying a yearly composition of 8s for his meadow ground in Langwathby Holm and as restoring the church porch at Great Salkeld. "Birch Close" is quite near to Nunwick Hall, being marked on the Tithe map of Great Salkeld Parish, 1840, in a position NY 556360.
56. Lees, *Flora of West Yorkshire*, page 158 refers this record to *Drosera anglica*.

	4. Folijs odoratis. Eglantine; or **Sweet-briar**. Nigh Spittle by KENDAL.	*Rosa rubiginosa* agg. Spital SD 527943
	Rosmarinum Sylvestre nostras. **Wild Rosemary**. T.L. BRIGSTEER-MOSS. Item On MIDDLETON-moss by LANCASTER; and in other Northern and SCOTCH mosses.	*Andromeda polifolia* L. Brigsteer village SD 482896 Middleton village SD 424587
	Rubia. **Madder**. 1. Sativa, Tinctorum.	
page 140	2. Minor pratensis coerulea. In the corn 'twixt SWARTHMOOR and ULVERSTON.	*Sheradia arvensis* L. Swarthmoor Hall SD 282773
	3. Angulosa aspera. Vide Mollugo. **S**	*Galium mollugo* L.
	4. Cynanchica. **Squinancy-wort**; or **Small English Saxifrage**. T.L. CONSWICK-SCAR.	*Asperula cynanchica* L. Cunswick Scar SD 491940
	5. Erecta Quadrifolia. T.L. Near SHAP-HEAD. [J.R.] **S** ORTON WINANDERMEER, et alibi in Westmorlandia. J.R.	*Galium boreale* L. Orton NY 623083
	Rubus. **Blackberry-bush**; or **Bramble**. 1. Major vulgaris; fructu nigro. **S**	*Rubus fruticosus* agg.
	2. Minor fructu Coeruleo. **S**	*Rubus caesius* L.
page 141	3. Alpinus Humilis; seu Chamaerubus Saxatilis. At HELS-FELL-WOOD; copiose.	*Rubus saxatilis* L. Helsfell Nab SD 503936
	4. Idaeus Spinosus. **Raspberry-bush**; **Framboise**, or **Hindberry**. **S**	*Rubus idaeus* L.
	Ruscus, vel Bruscus. **Kneeholm** or **Holly**; **Butchersbroom**. AS. Cneow-hole Aelf. Gloss. Near MILLUM-Castle.	*Ruscus aculeatus* L. Millom Castle SD 171814
	Ruta. **Rue**. 1. Muraria. Vide Adianthum Album. 2. Pratensis. Vide Thalictrum.	
page 142	**S**	
	Sagitta, seu Sagittaria. **Arrow-head**. 1. Major, Latifolia. 2. Minor, Latifolia. 3. Minor, Angustifolia.	

Saginae Spergula. Vide Spergula.

Salicornia. Vide Kali majus.

Salix. **Willow**

1. Alba vulgaris, Angustifolia. **S**	*Salix alba* L.
2. Folio Amygdalino, utrinque aurito, corticem abjiciens.	
3. Folio splendente Latifolia. \fragilis/ [T.L] SALKELD, EDENHAL, etc. **S** CRACK WILLOW.	*Salix fragilis* L. ? Little Salkeld Edenhall NY 565324

page 143

4. Folio splendente auriculato flexilis.	
5. Humilis, Corticem abjiciens.	
6. Folio Laureo, seu lato glabro odorato. [Qu. An diversa a *tertia* J.R. in montosis, septentrionalibus.] **S** SWEET-WILLOW or BAY-WILLOW.	*Salix pentandra* L.
7. Humilior folijs angustis subcoeruleis. **S** On the sides of the Caus-way from Grass-beck,[57] etc.	*Salix purpurea* L.
8. Pumila angustifolia.	
9. Pumila folio rotundo. On the top of INGLEBOROUGH and WHERNSIDE. KIRSKSTON.	*Salix herbacea* L. Whernside summit SD 738814 Kirkstone Pass NY 401081
10. Latifolia, inferne hirsuta. **Common Sallow. S**	*Salix cinerea* L.
11. Longissimis et Angustis folijs crispis subtus Albicantibus. **The Osier**. **S**	*Salix viminalis* L.

page 144

12. Longifolia non Auriculata, folio et viminibus subluteis. **Long-leav'd yellowish Sallow**.	

Salvia Agrestis; Sphacelus, seu Scorodonia. **Wood-sage. S** ON THE ROCKS NEAR THE MILL.[58] In Sylvis Arenosis, etc. passim.	*Teucrium scorodonia* L.

Sambucus. **Elder**; **Hiller**, or **Bore-tree**.

1. Vulgaris. **S**	*Sambucus nigra* L.
2. Fructu albo.	
3. Aquatica. **S**	*Viburnum opulus* L.

---

57. See above n 63 page 33.
58. See above n 35 page 65.

	4. Laciniata. T.L. near MANCHESTER.	*Sambucus nigra* L.
	5. Humilis. i.e. Ebulus.	
	Sanicula. **Sanicle. S**	*Sanicula europaea* L.
	Saponaria. **Sopewort**.	
	1. Vulgaris. About SEDBAR, ROSE, PLUMPTON, etc.	*Saponaria officinalis* L. Sedbergh Rose Castle NY 371461 Plumpton Wall NY 498371
	2. Concava Anglica.	
page 145	Satyrium. Vide Orchis.	
	Saxifraga. **Saxifrage**.	
	1. Alba vulgaris. **S** radice granulosa. In the woods and under shady hedges.	*Saxifraga granulata* L.
	2. Aurea. quae vel –	
	1. Rotundifolia, vulgaris. **S**	*Chrysosplenium oppositifolium* L.
	2. folijs pediculis oblongis insidentibus.	
	3. Anglica. Haecque varia –	
	1. Alsinefolia Annua. Pl. Oxf.	
	2. Graminea. **Pearl-wort. S** Caus-way in the HOLM[59] etc.	*Sagina procumbens* L.
	2. Alsinefolia folijs brevioribus, crassioribus et succulentioribus. T.L. in Fasc. On WHINNY-FIELD-BANK by CULLERCOATS near TINMOUTH.	*Sagina maritima* Don Whinny-field-bank Not located
	3. Folijs foeniculi latioribus **S** Meadow saxifrage.	*Silaum silaus* (L.) Schinz & Thell.
	4. Palustris. On CLIFTON-moor,[60] in West*morland*.	*Sagina nodosa* (L.) Fenzl
page 146	Scabiosa. **Scabious**.	
	1. Major vulgaris, flore laciniato. **S**	*Knautia arvensis* (L.) Coult.

---

59. See above n 11 page 10.

60. Clifton Moor was a roughly triangular area of unenclosed land, having its apex at Clifton Town End and its centre just under a mile south-easterly of this (NY 542249) about three hundred yards east of the A6. It became famous just after Nicolson's time as the scene of the last action fought on English soil when the retreating forces of Prince Charles Edward Stuart were engaged, on 18th December, 1745, in a skirmish with the Duke of Cumberland's pursuing army. (R. S. Ferguson, "The Retreat of the Highlanders through Westmorland in 1745", *CW1* x 186–228.)

2. Major vulgaris flore pleno.	
3. Minor; sive Columbaria. **S**	*Scabiosa columbaria* L.
4. Ovilla. Vide Rapunculus Scab*iosae* capitulo. **S**	*Jasione montana* L.
5. Radice succisa. i.e. Morsus Diaboli.	
Scandix. Vide Pecten veneris.	
Scolopendria. Vide Asplenium.	
Scrophularia. Johns.	
1. Major; sive Ocymastrum Alterum. **Brownwort**. **Figwort**. J.R. **S** called in some places of WESTMERLAND, HASTY ROGER.	*Scrophularia nodosa* L.
2. Minor. Vide Chelidonium minus.	
Scordium. **Water-Germander**. scordium alterum. Vide Salvia Agrestis.	

page 147

Secale. **Rye**. **S** Qu. WARE-RYE. An Species distincta.	*Secale cereale* L.
Sedum, **Sengreen**. Vide Saxifraga alba.	
1. Vulgare majus. **Great Houseleek**. **S**	*Sempervivum tectorum* L.
2. Minus. **Prick-madam**.	
1. Vulgare, flore luteo acuto.	
2. Acre flore luteo. **Wall-pepper**; or **Stone-crop**. **S**	*Sedum acre* L.
3. Officinarum, flore albo.	
4. A Rupe *Sancti* Vincentij. J.R.	
3. Alpinum. **Mountain Sengreen**.	
1. Trifido folio. [T.L.] Near LOWTHER, and BUCKBARROW. SHAP. MALHAM COVE. INGLEBOROUGH.	*Saxifraga hypnoides* L. Lowther NY 535236 Buckbarrow Crag NY 482075 Malham Cove SD 897641
2. Luteum minus. 'Twixt SHAP and ANNA-WELL, etc. INGLEBOROUGH.	*Saxifraga aizoides* L. Anna Well NY 584127
3. Purpureum ericoides. Ingleborough, etc.	*Saxifraga oppositifolia* L.
4. Minimum non Acre, flore albo. J.R. On the Rocks near WINANDERMEER. RYDALE-FELLS. All along the Coast from ST BEES to MILHAM.	*Sedum anglicum* Huds. St Bees Head NX 952133 Millom
5. Palustre parvum, flore incarnato. T.L.	*Sedum villosum* L.

page 148

on GATESCARTH-FELL.[61] HARTSIDE and INGLEBOROUGH. J.R. — Hartside Pass NY 647419

6. Tridactylites Tectorum. i.e. Paronychia folio Rutaceo.

[7. Ericoides coeruleum. INGLEBOROUGH.]

Selinum, Sij folijs. **Honewort**, or Corn-parsley.

Senecio. **Groundsell**; or **Herb Simpson**.

1. Vulgaris, minor. **S** — *Senecio vulgaris* L.
2. Major Hirsutus, foetidus. About SUNDERLAND nigh LANCASTER. — *Senecio viscosus* L. Sunderland SD 427557

Seriphium. Vide Absinthium Marinum.

Serapias. Vide Orchis palmata.

Serpentina. i.e. Coronopus Marina. Johns.

page 149

Serpyllum. **Wild Thyme**.

1. Vulgare. flore majore et minore. **S** — *Thymus drucei* Ronn.
2. Citrij odore. **Lemon-Thyme**.
3. Hirsutum Angustifolium repens. **S** On dry hills. — *Thymus drucei* Ronn.
4. Villosum fruticosius floribus dilute rubentibus.

Serratula. **Saw-wort**. **S** — *Serratula tinctoria* L.

Sesamoides Salamanticum magnum. **Spanish Catch-fly**.

Seseli Creticum minus. Vide Tordylium.

Sideritis.

1. Arvensis Rubra. Vide Ladanum Segetum.
2. Arvensis latifolia glabra. About WAKEFIELD, SHEFFIELD, etc. J.R. — *Galeopsis segetum* Neck.

---

61. I have not positively located this, but feel that Nicolson may have intended the fell slopes rising to Gatescarth Pass (NY 474094) as he appears to have botanised frequently in this area, possibly using the old drove road through Mosedale and down Longsleddale on his visits to Kendal and to Archer relations.

3. Anglica, radice strumosa. i.e. Marrubium Aquat*icum* Acutum. **Clown's Al-heal. S** — *Stachys palustris* L. — page 150
4. Humilis, lato folio. **Petty Al-heal. S** — *Stachys arvensis* (L.) L.
5. Hieraclia. Vide Stachys.

Sigillum B*eatae* Mariae. Vide Bryonia nigra.

Sigillum Solomonis. Vide Polygonatum.

Sinapi. **Mustard**.
1. Sativum vulgare. **S** — *Brassica nigra* (L.) Koch
2. Album. Vide Rapistrum Arvorum.
3. Sylvestre minus, Bursae pastoris folio. **Small wild mustard**. Johns.

Sison; sive Amomum officinarum. **Bastard Stone-parsley.** — page 151

Sisymbrium. Vide Mentha Aquatica.

Sium; sive Pastinaca Aquatica. **Water-Parsnep**, or **Cresses**.
1. Maximum Latifolium. **S** — *Sium latifolium* L.
2. Majus alterum, Angustifolium.
3. Minimum. T.L. Near MABURGH a bogg on the pasture. — *Apium inundatum* (L.) Rchb. f. Mayburgh[62] NY 519284
4. Umbellatum repens.

Smilax laevis. Vide Convolvulus major.

Smirnium. Vide Hipposelinum.

Solanum, **Nightshade**. — page 152
1. Vulgare, Officinarum. Under the walls at CARLILE. Item Under the Wall on the right hand of the way from MARSH-GRAINGE to DUNNERHOLM. — *Solanum nigrum* L. Marsh Grange[63] SD 220797 Dunnerholme SD 212798
2. Lethale; sive Bella Donna. ⟨**D**⟩**eadly Nightshade**, or **Dwale**. On and under the Walls at CARLILE; over against the — *Atropa bella-donna* L.

62. See above n 8 page 9.
63. See above n 14 page 11.

Abbey-Mill.[64] NEWBY-STONES.[65] CARTMEL-GRAINGE.[66] — Newby NY 593213

3. Lignosum; sive Dulcamara. **Bitter-sweet**. **S**
   1. Vulgare; in Sepibus. **Fellon-wood**, Nostratibus. **S** — *Solanum dulcamara* L.
   2. Marinum.
4. Tetraphyllon. Vide Herba Paris.

Soldanella Marina. **Sea-Bindweed**; **Sea-Colewort**, or **Scotch Scurvy-grass**. Isle of WALNEY, on the Sea-bank by TUMBERHILL. At ROOSBECK in FOURNESS. — *Calystegia soldanella* (L.) R. Br. Tummer Hill SD 179675 Roosebeck SD 258678

page 153

Sonchus. **Sow-thistle**.
1. Laevis vulgaris. **Smooth Sow-thistle**, or **Hares-Lettuce**. **S** — *Sonchus oleraceus* L.
2. Laevis Muralis. At MANNOR-Abbey[67] in FOURNESS. — *Mycelis muralis* (L.) Dum.
3. Asper; Laciniatus.
4. Arborescens. Great STRICKLAND. FOURNESS-Abbey. — *Sonchus arvensis* L. Great Strickland NY 560230 Furness Abbey SD 219717

Sophia Chirurgorum; sive Seriphium Germanicum. **Flixweed. S** — *Descurainia sophia* (L.) Webb ex Prantl

Sorbus. **The Service-Tree**.
1. Torminalis; sive vulgaris. T.L. In LEVENS-PARK. — *Sorbus torminalis* (L.) Crantz Levens Park SD 502855

page 154

2. Sylvestris, sive Aucuparia; quae et Ornus, sive Fraxinus Sylvestris. **Quicken-Tree**; **Roan-Tree**, or **Wild Service. S** — *Sorbus aucuparia* L.

---

64. A Map of the River Caldew and its Mills c.1775 (DB/13/55 in the Records Office, Carlisle) shows the Abbey Mill about a furlong south-westerly of the present Citadel Railway Station on the R. Caldew. See also above n 37 page 21.
65. For the name, see *Nicolson and Burn*, vol i, page 451.
66. See above n 3 page 6.
67. The Prestons of Holker, who acquired the site of Furness Abbey in the early 17th century, built their manor-house there. It was subsequently converted into a hotel (pages 224, 237–38, W. H. St John Hope, "The Abbey of St. Mary in Furness, Lancashire" *CW1* xvi 221–302.)

3. Aria Theophrasti; sive pilosa. **White Beam-Tree**, Chess-Apples, or Sea-Oulers. On METHOP Scarrs[68] over against MILTHROP. Westm*orland*. — *Sorbus aria* agg. Milnthorpe
4. Pyriformis. Pl. Staff. forte, Sativa. J.R. Fasc.

Sparganium. **Burr-reed**; or **Burr-Flag**.
1. Ramosum. **S** — *Sparganium erectum* L.
2. Non Ramosum. **S** — *Sparganium emersum* Rehm.
3. Minimum. **S** — *Sparganium minimum* Wallr. prob. includes *S. angustifolium* Michx.

Spartum. Vide Gramen Sparteum.

Spatula foetida. Vide Xyris.

Speculum Veneris minus. **The less Venus-looking-glass**; or **codded Corn-violet**. In the Corn by BARNBY ot'h' moor. T.L. — *Legousia hybrida* (L.) Delarb. Barnby Moor page 155

Spergula. **Spurry**.
1. Vulgaris, flore Albo. **S** Inter Segetes. — *Spergula arvensis* L.
2. Purpurea. **Purple Chickweed-Spurry**. **S** On the wall about KIRKFLATT.[69] — *Spergularia rubra* (L.) J. & C. Presl
3. Marina. On o*u*r Shores (Especially in the Isle of WALNEY) copiose. — *Spergularia media* (L.) C. Presl or *S. marina* (L.) Griseb.

Sphondylium. **Cow-parsnep**.
1. Vulgare Hirsutum **S** — *Heracleum sphondylium* L.
2. Folijs Laciniatis.

---

68. The limestone scar about a quarter of a mile south of Low Meathop village at SD 433792 is still an interesting botanical site.
69. Kirk-flatt or Curflatt is mentioned by Nicolson in his terrier of 1703–04, page 183, and a field in Great Salkeld at NY 553383 still bears the name. See also *CW2* xxv 118–19.

page 156

Spina. **A Thorn**.
1. Apendix. Vide Oxyacanthus.
2. Infectoria, seu Cervina. Vide Rhamnus.

Spongia. **Sponge.** ϵπόγγος, παρὰ τὸ σπᾶν τὰ ὑγρα.[70]
1. Ramosa Britannica.
2. Ramosa fluviatilis Newtoni.
3. Fistulosa Veneta.

Stachys major Germanica. **Base stinking Horehound.**

page 157

Staphylodendron. **The Bladder Nut-Tree**. Near. PONTEFRAICT.[71] — *Staphylea pinnata* L.

Stellaria Aquatica. **Water-Starwort**.
1. Folijs longiusculis. **S** In the ditch by the Caus-way from Grass-beck.[72] — *Callitriche* sp. prob. *C. platycarpa* Kütz.
2. Folijs longis tenuissimis.

Subularia lacustris. D*octor* LLoyd.

Succisa. Vide Morsus Diaboli.

Stratiotes. Vide Militaris Aizoides.

Symphytum. Vide Consolida major.

Synanchice. Vide Rubia Cynanchica.

T

page 158

Tanacetum, vulgare Luteum. **Tansie**. **S** All along the banks of EDEN, on both sides. — *Chrysanthemum vulgare* (L.) Bernh.

Tapsus Barbatus. Vide Verbascum.

Taxus. **The Yew-Tree**. In BRIG-STEER-Scars. Fells about RYDALL. — *Taxus baccata* L. Brigsteer village SD 482896

---

70. a sponge, because it draws in moisture.
71. Nicolson has taken this location from Ray who, although recording the tree from near Pontefract, was doubtful as to whether it was native (Raven, *Ray*, page 239 and Ray, *Historia Plantarum*, (1688), vol ii, page 1681.)
72. See above n 63 page 33.

Telephium; Crassula, seu Faba inversa. **Orpine**, or **Livelong**. **S** Roseum. i.e. Rhodia radix.	*Sedum telephium* L.	
Teucrium [pratense]. Vide Chamaedrys. [Sylvestris].		
Thalictrum. **Meadow-Rue**.		
1. Majus. Quibusdam, **Bastard Rubarb**. **S**	*Thalictrum flavum* L.	
2. Minus. **S** Sea-shore. MALHAM and SETTLE. J.R.	*Thalictrum minus* L. Malham village SD 901628	
3. Montanum Minus. Fasc.		
4. Foetidissimum. about WINANDER-MEER, in the great Isle. qu.	prob. *Thalictrum minus* L. Belle Isle SD 394967	
Thlaspi.		page 159
1. Dioscoridis, Drabae folio. **Treacle-Mustard**; or **Penny-cresse**. T.L. On LANCEMOOR. **S**	*Thlaspi arvense* L. Lansmere NY 575217	
2. Hederaceum. **Ivy-leav'd Mustard**.		
3. Montanum Glasti folio.		
4. Minus Angustifolium. **Bowyers Mustard**.		
5. Vaccariae incano folio perenne. **Perennial Mithridate-Mustard**.		
[6. Vasculo sublongo intorto].		
6. Vaccariae folio glabrum. By Mr LAWSON, in the Corn on LANCE-MOOR.	*Lepidium heterophyllum* Benth. Lansmere NY 575217	
7. Vulgatissimum, folio vaccariae. **Mithridate-Mustard**: or **Bastard Cresses. S**	*Lepidium campestre* (L.) R.Br.	
Tilia foemina. **The Lime**, **Line** or **Linden-Tree**.		page 160
1. Vulgaris Major, platyphyllos. Ladyholm in WINANDER-MEER. BRIGSTEER.	*Tilia platyphyllos* Scop. Lady Holme SD 398975 Brigsteer SD 482896	
2. Minor, folio minore. Quibusdam (a funibus) Bast.		
Tithymalus, **Spurge**.		
1. Characias vulgaris. **Wood-spurge**.		
2. Helioscopius. **Sun-Spurge. S**	*Euphorbia helioscopia* L.	
3. Hibernicus. **Makinboy**; or **Knotty-rooted Spurge**.		

	4. Leptophyllos. Vide Esula Exigua.	
	5. Segetum Longifolius. N.B. **S** WILLIES-WOOD,[73] by the water side.	*Euphorbia esula* L.
page 161	5. Maritimus. Sea-spurge. I*sle* of WALNEY.	*Euphorbia paralias* L.
	6. Pineus. Ger. **S** NB.	*Euphorbia esula* L.
	6. platyphyllos.	
	7. Peplus dictus, folijs subrotundis.	
	Tordylium.	
	Tormentilla. **Tormentil**; or **Septfoil**.	
	1. Vulgaris, Officinarum. **S**	*Potentilla erecta* (L.) Räusch.
	2. Flore pleno. T.L. At TEMPLE-SOWERBY.	*Potentilla erecta* (L.) Räusch. Temple Sowerby NY 611272
	3. Reptans alata, folijs profundius seratis. Pl. Oxf.	
	Trachelium. **Throatwort**; or **Canterbury-Bells**.	
	1. Majus Belgarum, seu Giganteum. T.L. At GREAT STRICKLAND. **S**	*Campanula latifolia* L. Great Strickland NY 560230
	2. Urticae folio. **S**	*Campanula trachelium* L.
	3. Minus. **S**	*Campanula glomerata* L.
page 162	Tragon. Vide Kali Spinosum.	
	Tragopogon. **Goatsbeard**; or **Goe-to-bed-at-noon**.	
	1. Luteum. T.L. At GREAT STRICKLAND. **S**	*Tragopogon pratensis* L. Great Strickland NY 560230
	2. Purpureum. Qu. J.R. Anglicum *non* credit. About ROSE. Item in the fields about CARLILE.	*Tragopogon porrifolius* L. Rose Castle NY 371461
	Tragopyron. Vide Fegopyron.	
	Tribulus Aquaticus minor. **Small Frogs-Lettuce**; or **Water-Caltrops**.	
	1. Muscatellae floribus.	
	2. Quercus floribus. **S**	*Potamogeton crispus* L.

73. See above n 44 page 23.

Entry	Identification	Page
Trichomanes. Vide Capillus veneris. **English black Maiden-hair**. J.R. **S** Ramosum.	*Asplenium trichomanes* L.	
Trifolium. **Trefoil**.		page 163
1. Acetosum vulgare; sive Alleluia Officinarum. **Wood-sorrel. S**	*Oxalis acetosella* L.	
2. Cochleatum folio cordato maculato. **Heart-Trefoil**; or **Claver**.		
3. Echinatum. **Hedge-Hog-Trefoil**.		
1. Arvense, parvum.		
2. Cochleatum modiolis Spinosis.		
4. Corniculatum. **Bird's-foot-Trefoil**. Majus,[74] **Minus** et **Minimum**. **S**		
5. Fragiferum. **Strawberry-Trefoil. S**	*Trifolium fragiferum* L.	
6. Lupulinum. **Hop-Trefoil**. Majus et **Minus**. **S**	*Trifolium dubium* Sibth.	
7. Pratense. Quod vel –		page 164
1. Album. Hocque –		
1. Minus (et foemina) glabrum.		
2. Folijs purpureis.		
3. Umbella siliquosa.		
4. Hirsutum majus, flore Albosulphureo. **S**	*Trifolium ochroleucon* Huds.	
2. Purpureum. Hocque –		
1. Vulgare. **Honey-suckle-Trefoil**. **S**	*Trifolium pratense* L.	
2. Majus, seu Sativum. Claver-grass. Vide tamen Fasc.		
3. Glomerulis florum oblongis.		
4. Glomerulis mollioribus et rotundis.		
3. Luteum, \odoratum./ **Melilot**. **S** Vide Melilotus.	*Medicago lupulina* L.	
4. Stellatum Glabrum.		page 165
5. Pumilum Supinum, flosculis longis Albis.		

74. "Trifolium corniculatum majus" which Nicolson does not underline and therefore, presumably, does not consider that he has seen (see page 59), is *Lotus pedunculatus*. "Minus" would then be *L. corniculatus* and the most likely identification for "minimum" would be *L. tenuis*. In view, however, of the fact that Nicolson appears only to have distinguished two species, it seems more likely that he has assigned his specimens to the wrong Raian synonyms – his "minus" to *L. pedunculatus* and his "minimum" to *L. corniculatus*.

8. Palustre; sive Paludosum. **Marsh-Trefoil**; or **Buck-beans**. **S** — *Menyanthes trifoliata* L.
9. Sylvestre Luteum. Sive Medica Sylvestris.

Tripolium. Majus et MINUS. **Sea Starwort**; **Blew Sea-Daisies**; or **Hoggs-beans**. 'Twixt CARTMEL-WELL[76] and GRAINGE. Isle of WALNEY. MIDDLETON near LANCASTER. — *Aster tripolium* L.[75] Grange-over-Sands Middleton SD 424587

Triticum. **Wheat.**
1. Album, spica mutica.
2. Rubrum.
   1. Spica mutica albicante, granis rufescentibus.

page 166

   2. Spica Aristata, granis et Spica Rufescentibus.
   3. Spica Aristata Albicante, granis solum rufescentibus.
3. Aristis circumvallatum, glumis hirsutis. **Gray wheat**; **Duck-bill wheat**; or **Gray pollard**.

Turritis, **Tower-mustard**. T.L. In CLIBURN. **S** — *Turritis glabra* L. Cliburn NY 588245

Tussilago. **Colts-foot**; or **Foals-foot**. **S** — *Tussilago farfara* L.

Typha. **Catstail**; or **Reed-mace**.
1. Major. [T.L.] In BARWICK-HOLM near HORNBY.[77] HAWKESHEAD. — *Typha latifolia* L.
2. Minor.

page 167

V

Vaccinia.
1. Nigra. Quae vel –
   1. Vulgaria. **Black whorts**; **Whortle-Berries**; or **Bill-berries**. **S** — *Vaccinium myrtillus* L.

---

75. The form "minus" is probably the form lacking the purple ray-florets (See Gilmour, *Botanical Journeys*, page 61). See also above n 74 page 89.
76. See above n 13 page 11.
77. Not located, unless this is one of the two small islands in the R. Eamont south-westerly of Hornby Hall and shown on the Tithe map of Brougham Parish, 1839. The field adjoining the river here is called Barwick Bank (NY 562295). See also above n 11 page 10.

2. Fructu majore. T.L. In WHINFIELD[78] J.R. **S** Troughfoot in LAZONBY parish, RENWICK-Holms,[79] etc. — *Vaccinium uliginosum* L. Whinfell Forest NY 573275 Troughfoot NY 469425

2. Rubra. Haec itidem –
   1. Folijs buxeis. In WHINFIELD.[78] BARON-WOOD. etc. — *Vaccinium vitis-idaea* L. Whinfell Forest NY 573275 Baron Wood NY 510440
   2. Folijs myrtinis crispis.
3. Nubis. Vide Chamaemorus.
4. Palustria. **Marsh-whortleberries**; **Moss-berries** or **Moor-berries**. Nostras **Crone-berries**. In CLIBBURN-MOSS. — *Vaccinium oxycoccos* L. Cliburn Moss NY 577257

Valeriana; sive Phu. **Valerian**.

page 168

Vide Lactuca Agnina.

1. Sylvestris magna Aquatica. **S** — *Valeriana officinalis* L.
2. Palustris minor. **S** — *Valeriana dioica* L.
3. Minor altera. Floribus priori quintuplo minoribus.
4. Graeca; sive coerulea. **Jacob's-Ladder**. On the sides of INGLEBOROUGH.+ A⟨t⟩ MALHAM-COVE and the WHERN.[80] Near BRIGNEL-Church among the Shrubs, etc. +Near DALEMAIN.[81] — *Polemonium caeruleum* L. Malham Cove SD 897641 Brignall Church NZ 073123 Dalemain NY 477269
5. Minima Lactucae Agninae similis.

Verbascum. **Mullein**.

1. Latifolium Luteum; seu Tapsus barbatus. **White Mullein**; **High Taper**; or **Cows Lungwort**. On the old walls at KENDAL-CASTLE, etc. — *Verbascum thapsus* L. Kendal Castle SD 522924

page 169

2. Nigrum, flore ex lut⟨eo⟩ purpurascente. **Black Mullein**.
3. Flore albo parvo.

---

78. *PNW* ii 132.
79. See above n 25 page 14.
80. This is a location of John Ray, who describes the plant as growing at "Cordil or the Wern" (*Synopsis*, 1724, page 288). Gordale Beck is a mile and a half east of Malham Cove.
81. In Nicolson's text this is entered as a side note: it is his own location for the plant, whereas the others are taken from Ray.

4. Pulverulentum, flore luteo parvo. **Hoary Mullein**.

Verbena. **Vervain**. COCKERMOTH-Castle. LEVENS-Mill. — *Verbena officinalis* L. Cockermouth Castle NY 125310

Vermicularis frutex \marina./ **Shrub**, or **Tree-Stonecrop**.
1. Major.
2. Minor.

Veronica. **Speedwell**; or **Fluellin**. Vide Anagallis Aquatica. Alsine veronicae folio. Chamaedrys Sylv*estris*.

page 170
1. Foemina. Vide Elatine.
2. Mas, vulgaris. **S** — *Veronica officinalis* L.
3. Minor pratensis. **Paul's-Betony**.
4. Spicata. Quae vel –
   1. Major. On the rock above CARTMELL-Well. J.F.R. — *Veronica spicata ssp. hybrida* (L.) E.F. Warb.[82]
   2. Minor.

Vesicaria Marina. forte Ostrearum foetura. J.R.

Viburnum; seu Lantana. **The Wayfareing-Tree**; or **Cotton-Tree.**

page 171
Vicia. **Vetch**; or **Tare**.
1. Vulgaris Sativa, semine nigro.
2. Sylvestris major. clavientata. **S** among the Corn. — *Vicia sativa* L.
3. Multiflora Dumetorum. **S** — possibly *Vicia cracca* L.
4. Multiflora maxima. In the woods about GREATA-BRIDGE. J.R. At the bridge end at KIRKBY-LAUNSDALE. In Scarbank 'twixt CAMMERTON and WORKINTON. — *Vicia sylvatica* L. Greta Bridge NZ 086132 The Devil's Bridge SD 616783 Scarbank Wood NY 026305
5. Sepium, perennis. **Bush-Vetch**. **S** — *Vicia sepium* L.
6. Flore luteo, siliqua Hirsuta.

---

82. Cartmel well on Humphrey Head (See above n 13 page 11) is still an important location for this rare plant. For a note on John Fitz-Roberts, see page xxxii.

7. Lathyroides, purpuro-coeruleis floribus. EDINBURGH-Park.[83]	*Lathyrus palustris* L.	
8. Segetum cum siliquis multis Hirsutis. **S** Nostras TRADDAH.	*Vicia hirsuta* (L.) S.F. Gray	
9. Minima, cum siliquis glabris. **S**	*Vicia tetrasperma* (L.) Schreb.	
Vinca pervinca. Vide Clematis Daphnoides.		page 172
Viola. **A Violet**. AS. Apple-leaf.		
1. Aquatica. Vide Millefolium Aquat*icum*.		
2. Martia. **Common violet**. quae –		
1. Alba. **S**	*Viola odorata* L.	
2. Purpurea. **S**	*Viola odorata* L.	
3. Inodora Sylvestris. **Wild** or **Dog's violet. S**	*Viola canina* L.	
4. Hirsuta inodora. Pl. Oxf.		
3. Bicolor Arvensis. **Small wild Hearts-Ease**; or **Pansies. S**	prob. *Viola arvensis* Murr.	
3. Palustris rotundifolia glabra.		
3. Tricolor major. **Common Hea\r/ts-Ease. S**	*Viola tricolor* L.	
4. Montana Lutea grandiflora. BOWES. On the mountains about MALHAM in CRAVEN. Near Dumma-hil (on Penyston-green) juxta BURGH under STAINMOOR.	*Viola lutea* Huds. Dummah Hill Malham village SD 901628 Penistone Green NY 835156 Brough	
5. Rubra striata Eboracensis. Fasc.		
Viorna. **Great wild Climber**; or **Travellers joy**. L.[84]	*Clematis vitalba* L.	page 173
Virga Aurea. **Golden-Rod**.		
1. Vulgaris. T.L. In CLIBURN-FIELD. **S**	*Solidago virgaurea* L. Cliburn village NY 588249	
2. Folijs serratis. **S**	*Senecio fluviatilis* Wallr.	
3. Montana; folijs angustis subincanis, flosculis conglobatis.		
Virga pastoris. Vide Dipsacus minor.		

---

83. See above n 6 page 7.
84. See above n 27 page 15.

Viscum. **Misseltoe**. T.L. About BRIGSTEER-MOSSE. AS. Ac-mistel. Widow Moor's son at BRIGSTEER will find it. — *Viscum album* L. Brigsteer village SD 482896

Vitis Idaea [pal*ustria*]. Vide Vaccinia [palustria].

Ulmaria. Regina prati. Barba Capri. **Meadow-sweet**. **S** — *Filipendula ulmaria* (L.) Maxim.

Ulmus. **An Elm-Tree**. **S** — *Ulmus procera* Salisb.

page 174
1. Vulgaris, folio lato scabro. [**S**].
2. Folio latiore. **Broad-leav'd Elm**; or **Wych-Hasell**.
3. Folio glabro. **Wych-Elm**.
4. Minor, folio angusto scabro.

Umbilicus Veneris. **Wall-pennywort**; **Navel-wort**; or **Kidneywort**. [**S**]

Urtica. A Nettle. **S**
1. Major Urens. **S** — *Urtica dioica* L.
2. Minor.
3. Romana. Sive Mas cum globulis.
4. Mortua. Vide Lamium.
5. Herculea. Vide Galeopsis.

page 175
Vulvaria. Vide Atriplex canina.

Vvularia. Vide Trachelium.

X

Xanthium. Vide Bardana minor.[85]

Xyris; sive Spatula foetida. **Stinking Gladdon**, or **Gladwyn**.

---

85. Altered from "major".

Index nominum Anglicorum omnium quae per paginas praecedentes sunt Dispersa.[86]

86. The page numbers in this list are Nicolson's. The index, which is far from complete, ends the *Catalogue of Plants*. Nicolson, however, continued to use the notebook to make occasional entries which are detailed in full or in part in the pages succeeding the index.

Miscellanea Northymbrica. 1695[87]

Plantae quae sequuntur Northymbricae merito dicendae *sunt* utpote quae (teste J. Raio) vel solummodo, vel praecipue sultem, intra veteris Northymbriae regni limites inveniuntur.

Acetosa rotundifolia repens Eborac*ensis*. "Acetosa ... Westmorlandica"
Adianthum petraeum perpusillum foliis bif*idis*.
—— 'Ακρόστιχον[88]
—— Album crispum alpinum.
—— Album floridum.
Alchimilla Alpina pentaphyllos.
Allium montanum bicorne purp*ureum* prolif*erum*.
—— Amphicarpon, folijs porraceis etc.
Apium marinum Scoticum.
Asphodelus Lancastriae.
—— palustris Scoticus minimus.
Bifolium minimum.
Bistorta minor nostras.
Buphthalmum vulgare.
Bursa pastoris majori affinis, loculo oblongo.
Cardamine impatiens. [...]
Carum.
+Cerasus fructu minimo cordiformi.
—— fr*uctu* parvo serotino.
—— Avium racemosa.
Chamaecistus montanus folio pilosellae min*oris*.
Chamáemorus.
Chamaepericlymenum.
Chamaerubus saxatilis. "Rubus Alpinus Humilis"
Christophoriana vulgaris.
Cirsium Britannicum Clusij. *Cirsium heterophyllum* (L.) Hill
Cochlearia major rotundifolia.
+—— folio anguloso parvo.

---

87. This list of plants which Nicolson takes on Ray's authority as being Northumbrian in the sense of being found either exclusively or chiefly within the ancient borders of that kingdom is substantially the same as one which is to be found on Bodleian Lib, MS Top Gen C 27, amongst the material that Nicolson was collecting for his history of Northumbria (see above page xxiv). Most of the plants listed are to be found in his own *Catalogue*, often with locations in Cumbria. Some, however, appear under a different polynomial and, where this is so, I have cross-referenced. To the 7 plants which do not appear in the *Catalogue*, or appear but without location, I have given modern names.
88. forked.

Cotyledon Hirsuta.
Crithmum spinosum.
Echium marinum.
Elaeagnus cordi.
Eruca monensis laciniata.
Euphrasia rubra Westmorlandica.
Filix, Saxatilis *non* ramosa nigris mac*ulis* punct*ata*.
—— Florida.
—— Saxatilis caule tenui fragili.
Fucus membranaceus ceranoides. *Rhodymenia palmata* (L.) Greville
—— Tinctorius.
—— Balteiformis.
—— Phasganoides.
—— Pols⟨c⟩hides.
Fumaria alba latifolia.
Fungus piperatus albus lacteo succo turgens.
Geranium moschatum. "Geranium Moschatum cicutae folio"
+—— Haematodes Lancastrense.
—— Batrachoides montanum nostras.
—— Columbinum maximum folijs dissectis. *Geranium columbinum* L.
Gladiolus lacustris.
Gramen junceum leucanthemum.
+—— Sparteum, capite bifido. *Isolepis setacea* (L.) R. Br.
Helleborine flore Atro-rubente.
—— minor alba.
+Hieracium macrocaulon.
—— *λεπτόκαυλον*.
—— fruticosum latifolium glabrum.
Hipposelinum.
Juniperus Alpina minor.
Leucoium vasculo sublongo intorto.
Lilium convallium Angustifolium.
Lunaria minor. et ramosa. item folijs dissectis. "Lunaria minor"
Lysimachia siliq*uosa* glabra minor latifolia.
—— lutea flore globoso.
Meum.
Muscus clavatus, folijs cupressi.
—— —— terrestris repens.
Muscus marinus, rubens pennatus.
—— —— Equisetiformis *non* ramosus.
Muscus Terrestris, polyspermos.
—— Erectus abietiformis.
—— Trichomanoides.
Nasturtium Petraeum.
Ornithogalum luteum.
Orobus Sylvaticus nostras.
Pentaphylloides fruticosum.
Persicaria siliquosa.

Pinus sylvestris conis parvis albentibus. qu. an intra regnum North*umbriae*. *Pinus sylvestris* L.
Polygonatum flore majore odoro.
Pyrola folio mucronato serrato.
—— Alsines flore {Europaea. / Brasiliana.}
[+Ptarmica flore pleno]
Raphanus rusticanus.
Rhodia radix.
Ribes f[lore] \ructu/ rubro vulgaris.
—— Alpinus dulcis. *Ribes alpinum* L.
Rorella longifolia perennis. "Ros solis Perennis longifolia"
—— longifolia maxima. "Ros solis Longifolia maxima"
Rubia erecta quadrifolia.
Salix pumila montana folio rotundo.
—— latifolia folio splendente.
Sambucus folijs laciniatis.
Saxifraga Alsinefolia fol*ijs* succulentioribus.
Sedum palustre parvum.
—— Alpinum {Ericoides purpureum. / minus luteum nostras. / trifido folio.}
Sideritis Arvensis latifolia glabra.
Sorbus Aucuparia.
Thalictrum minus.
+Thlaspi vaccariae folio glabrum.
Vaccinia nigra fructu majori.
—— rubra buxeis folijs.
Valeriana Graeca.
Vicia multiflora maxima.
Viola rubra striata Eboracensis. prob. *Viola palustris* L.
—— montana lutea grandiflora nostras.

Plantae quae cum + notantur, novae sunt et a T. *Lawson* primo observatae et D. *Ray* communicatae.[89]

89. A long list of northern plants was sent by Lawson to Ray in 1688 (see above page xl), some of which are included in the above list. Nicolson has marked with a cross those singled out as particular discoveries of Lawson. Some of these subsequently came to be regarded simply as varieties of plants which were already known.

His Addendae sunt quae ad marginem Catalogi *Joh. Raij* notantur manu propria *Tho. Lawson*[90]

(1) Alsine Becabungae folio, *morisoni*. In an old Ditch-stead 'twixt *Marsh-Grange* and *Dunnerholm* in *Furness*. Item On the sea-bank 'twixt *Bare* and *Pulton* nigh *Lancaster*. (*Samolus valerandi* L.)

(2) +Anonis Spinosa. flore albo. By *Bigger* in the Isle of *Walney*. (*Ononis spinosa* L.)

(3) Armerius pratensis flore albo. Near *Orton* in *Westmorland*. (*Lychnis flos-cuculi* L.)

(4) +Bifolium palustre, tribus folijs. In the low Hagg over against the Mill at great *Strickland*.[91]

(5) Cannabis Spuria flore albo magno Elegante. *Mer*. Every where near *Nottingham*. (*Galeopsis tetrahit* agg.)

(6) +Cardamine flore pleno. On little *Strickland*-pasture. Item about *Middleton* by *Lancaster;* some with 5 rowes of leaves. (*Cardamine pratensis* L.)

(7) +Cariophyllata flore amplo purpureo pleno. In Mr Lawson's Oxen-close and Gill at great *Strickland*. Plantam hanc sic ipse describit: Caryophyllata purpurea prolifera, quadruplici aut quintuplici serie petalorum; e medio floris emergit caulis florem in summitate gerens; flos ex 18 petalis constat. (*Geum rivale* L.)

(8) Cotula non foetida, flore pleno. Quadruplici aut Quintuplici serie petalorum. At great *Strickland*. (*Tripleurospermum maritimum* (L.) Koch ssp. *inodorum* (L.) Hyl. ex Vaarama)

(9) Equisetum nudum variegatum minus. J.B. at great *Salkeld*. Near *Thrimby* Chappel. Aphyllum. Aphyllocaulon. (*Equisetum arvense* L.)[92]

(10) +Geranium Batrachoides flore eleganter variegato. In old Deer-park by *Thornthwaite*.[93]

(11) Geranium Batrachoides longius radicatum, odoratum. Upon a wall by *Thornthwaite* Hall. Sponte an ab ejectamentis horti incertum.[94]

---

90. An account of these marginalia which were added by Lawson to his copy of John Ray's *Catalogus Plantarum Angliae* will be found on pages xl–xlvi.

91. This, as Martindale suggests (*The Westmorland Natural History Record*, vol i, page 119), is probably a three-leaved twayblade (*Listera ovata*, Ray's "Bifolium sylvestre"), rather than the Little Bog Orchid (*Hammarbya paludosa*) to which Ray had given the name "Bifolium palustre".

92. See Martindale, *Ibid*, pages 181–82.

93. See above n 76 page 39.

94. Martindale, *Ibid*., page 77 identifies this as a garden escape, *Geranium macrorrhizum*, suggesting that the garden it may have originated from was Sir John Lowther's.

(12) Geranium Columbinum folio Malvae rotundo, flore albo. Under a wall by the *Round-Table*. (*Geranium molle* L.)

(13) +Geranium Haematodes album, venis rubentibus striatum. Near *Bigger* in the Isle of *Walney*. (*Geranium sanguineum* L. var. *Lancastrense* (With.) Druce)

(14) Hesperis pannonica inodora. J.B. Park. Sylvestris inodora. *Wild Dames violets*. In the beck that parts *Yorkshire* and *Lancashire*, in the way from *Westby* in *Craven* to *Pendle-Hill*.

(15) Narcissus flore albo et albido. About *Ulverston* in *Lancashire*. Item Great *Strickland*. (*Narcissus pseudonarcissus* L.)

(16) Orchis palmata palustris Dracontias. Park. Near Common-holm-bridge and the mill at great *Strickland*.

(17) —— Cynosorchis militaris purpurea odorata. Purple Sweet Soldier's Cullions. Park. In Linham at *Blackwell* by *Tees*. On *Lance-moor* near *Newby* in West*morland* all about the Fairy-holes. (*Anacamptis pyramidalis* (L.) Rich)[95]

(18) Pedicularis\pratensis/ flore albo. At *Gunnerthwait* in *Lancashire*. (*Pedicularis sylvatica* L.)

(19) +Pedicularis palustris elatior alba. In the foot of long *Sleddale*. (*Pedicularis palustris* L.)

(20) +Ptarmica flore pleno. In the small holm in *Winandermeer*. (*Achillea ptarmica* L.)

(21) Rosa Sylv*estris* inodora sive Canina flore pleno. Near *Malham* in *Craven*. (*Rosa canina* agg.)

(22) +Saponaria flore pleno. At *Carnforth* in *Lancashire*. (*Saponaria officinalis* L.)

(23) Scabiosa montana maxima. *Lob*. Scabiosa Alpina Centauroides. *Cam*. Found at *Lowther*: but in a place which, he supposes, was once a garden.[96]

(24) Serratula folijs ad summitatem usque indivisis. *Reg. Harrison*. (*Serratula tinctoria* L.)

(25) +Thlaspi veronicae folio. *Parkins*. Bursae pastoriae loculo sublongo affinis pulchra planta. J.B. On the sides of the mountains in *Craven*. Its leaves lying next the ground are rough, hairy, almost round, indented; of a deep green colour, each one upon a short foot-stalk, somew*ha*t resembling the leaves of *Speedwell*, its stalk is hairy, half a foot high, branching usually from the bottom, sometimes without branches, at the top are many small white flowers, small round long pods, one above another spike-fashion, containing in each a small brownish yellow seed, Root white and long. (This Plant is in the former Catalogue. I only repeat it here for *Mr Lawson's* Description.) (*Draba muralis* L.)

---

95. So identified by Martindale, *Ibid*. page 153.
96. This is again a garden escape. Martindale, *Ibid*., page 182, identifies it as probably *Cephalaria alpina* Schrad.

(26) +Thlaspi folijs globulariae. J.B. On the pastures above the ebbing and flowing well near *Gigleswick*. In the mountainous pastures between *Settle* and *Malham*, varijs in locis. This has many small leaves lying on the ground round about the root, like the blew *Daisie,* full of juice, of a dark blewish green colour, its diverse stalks are about a foot high, bearing thereon many leaves, longer and more pointed than the lower, sundry white flowers at the topp one above another, bearing after flatt powches as *Shepherd's purse,* its root is long, white and fibrous. (This *Mr Ray* in his *Synopsis* has enter'd among his *English* plants: but takes no notice of his haveing it from Mr *Lawson*.) (*Thlaspi alpestre* L.)

(27) Thlaspi minus Clusij. Ger. 268. In the pastures above the ebbing and flowing well near *Gigleswick,* in stony ground, among the grass. Thlaspi perfoliatum minus. Mer. This hath a few leaves lying on the ground, of a greyish green colour, those that grow higher upon the stalk, which is scarce a foot high, and but a few sett thereon, are smaller, pointed at the ends, and broad at the bottoms, compassing it, the flowers are small and white, the seed-vessels flatt, some*what* sharp-pointed. (*Thlaspi perfoliatum* L.)

(28) Tormentilla flore pleno. At *Temple-Sowerby* in *Westmerland.* (*Potentilla erecta* (L.) Rausch)

(29) Tragopogon purpureum. In the fields about *Carlile. Ray* does *not* believe it. *Synop.* p. 47. (*Tragopogon porrifolius* L.)

(30) +Valeriana Graeca, flore albo. About *Malham*-Cove, abundantly. (*Polemonium caeruleum* L.)

(31) Virga Aurea latifolia serrata. C.B. It growes as plentifully in *ou*r fields at Salkeld as the *vulgaris.* (*Senecio fluviatilis* Wallr.)

. . . .

Those mark'd with an + are in Mr Ray's Synopsis, A.D. 1690. The Rest communicated to Cous*in* Archer. May. 14. 94.

. . . .

Tithymalus Pineus et Segetum longifolius. In the *Willies*-wood at *Salkeld.* (*Euphorbia esula* L.)

[Lists and Notes]

There follow some miscellaneous lists, compiled by Nicolson and relating to matters of Natural History. These have not been reproduced, as they are mostly derivative and do not deal with plants or botany. The headings of these lists and explanatory notes will be found below.

Heading: Of the Figure and colour of the Eggs of Birds.
Pages: One.
Notes: An attempt at a key to birds' eggs, working on the basis of shape and colour. There are references to Nehemiah Grew, *Musaeum Regalis Societatis*, (1681), page 75 and to John Ray's English edition of Francis Willughby's *Ornithology*, (1678), page 10.

Heading: *Musaeum Tradescantianum.* 8°. *Lond.* 1656. What in it relates to the Rarities of the Province of York, or antient K*ingdom* of Northumberland.
Pages: Two.
Notes: The title of John Tradescant's catalogue of his museum is: *Musaeum Tradescantianum: or, A Collection of Rarities Preserved At South-Lambeth neer London,* (London, 1656).
This collection of natural curiosities and rarities which was made by John Tradescant (1608–62) was one of the sights of London from the 1630s on. It was Britain's first public museum and became the foundation of the Ashmolean Museum when Elias Ashmole acquired it after Tradescant's death. For an account of it, see Mea Allan, *The Tradescants, their plants, gardens and museum 1570–1662*, (London, 1964), where the catalogue is reproduced on pages 247–312.

Heading: The like out of Grew's *Musaeum.* R.S. Fol. Lond. 1681.
Pages: Eight.
Notes: These are extracts made by Nicolson from Nehemiah Grew, *Musaeum Regalis Societatis or a Catalogue and Description of the Natural and Artificial Rarities Belonging to the Royal Society and preserved at Gresham Colledge ...,* (London, 1681).

Heading: The like out of *Mr Ray*'s Engl. Edition of *Mr Willughby's* Ornithology. Fol. Lond. 1678.
Pages: Seven.
Notes: These are extracts relating to birds seen in the North taken from Francis Willughby, *The Ornithology ... Translated into English, and enlarged with many Additions throughout the whole Work ...*

by John Ray, (London, 1678). An account of Ray's editorial work will be found in Charles E. Raven, *John Ray*, (Cambridge, 1942), pages 308–38. Nicolson prefaces his extracts with a reference to the Yorkshiremen Francis Jessop and Ralph Johnson, from whom Ray and Willughby received much information about Northern birds.

Heading: Musaeum Greshamense.
Pages: One.
Notes: This is a bare list of natural and artificial categories: "Fish", "Sea-plants", "Coins" etc., with no elaboration. It appears to have been compiled from the contents pages ("A Prospect of the whole Work of the Musaeum", A5^{v} – A6^{r}) of Grew's *Musaeum Regalis Societatis* (see above).

Plants desir'd by Mr Ja. Sutherland, in a List given me by him at Edenb*urgh* Jun. 16. 1701.[97]

Adianthum petraeum perpusillum folio bifido. A dozen good large Specimens.
Bifolium minimum. Sets and Specimens.
Bursa pastoris loculo oblongo. Seeds and Specimens.
Cerasus fructu Cordiformi. } Sets and ripe fruit.
—— fr*uctu* parvo serotino. }
Chamaecistus montanus folio pilosellae. Sets, Specimens and Seeds.
Chamaeperyclymenum. Sets, Specimens and Fruits.
Gramen Sparteum capite bifido. Sets, Seed and Specimens.
Hieracium macrocaulon. Sets, Seeds and Specimen. It*em*.
—— Leptocaulon et Fruticosum latifol*ium* glabr*um*.
Juniperus Alpina minor.
Leucoium Vasculo Sublongo intorto.
Persicaria Siliquosa.
Pyrola folio mucronato serrato.
—— Alsines flore Brasiliana.
Nasturtium petraeum.
Polygonatum flore majore odoro.
Ribes dulcis Alpinus.
Rorella longifolia perennis.
—— Longifolia maxima.
Salicum Species variae.
Saxifraga Alsinefolia folijs succulentioribus.
Sideritis Arvensis latifolia glabra.
Vaccinia Nigra fructu majore.
Viola rubra striata Eboracensis

Of all Sets, Seeds, Fruit and Specimens.

---

97. If Nicolson ever sent specimens up to Sutherland in Edinburgh, they are not now to be found amongst the herbarium specimens at the Royal Botanic Garden, Inverleith.

Specimens of some plants collected in Ireland by Mr Lhwyd, and sent by him to Mr Sutherland. 1701.

Adianthum folijs bifidis vel trifidis Newtoni.
Muscus Trichoma*noides* Aquatico-Alpinus. R. Synops.
Sanicula Alpina inter Guttatam et Sedum Serratum ambigua.
Erica Cantabrica flore max*imo* folijs Myrti subincanis. Tournfortij.
Adianthum folijs Coriandri. C.B.
Caryophyllus Arvensis polygoni folio, majore flore, Glaber. Sch. Bot. par.
Lonchitis Aspera major. B.P.
Caryophyllata Alpina Chamaedryos folio duriore. Morisoni.
Quinquefolium Alpinum Argenteum Album. Raij Catal. Exot.
Gentianula quae Hippion. J.B.
Vaccinia Rubra fol*ijs* myritinis Crispis. Merret.
Spongiola Utriculata compressa Alba.
Rorellae Species quarta. Raij Hist.
Millefolium palustre galericulatum minus. Raij Synops.
Trichomanes montan*um* Costa viridi quandoque bifida.
Geum folio circinato acute crenato pistillo floris rubro. Tournf.
Pentaphylloides fruticosum. Raij.
Sedum Serratum folijs pediculis oblongis insidentibus.
Arbutus folio serrato.

[Lists and Notes]

There follow further lists and notes relating to matters of Natural History. Headings and brief explanatory notes follow as before.

Heading: Catalogue of Sea-Fishes caught at Whitehaven. As in a Letter from Eb. Gales d*ated* Sep. 19. 1701, sent to S*i*r R. Sibbald, Sep. 23. 1701.
Pages: One.
Notes: A list of 30 different species.

Heading: Sea-Fish on the Coast of *Galloway* in Scotland, as by Letter from S*i*r R. Sibbald. Sep. 16. 1701.
Pages: Three.
Notes: 2½ pages deal with the sea fish, listing 28 different species with some notes. This is followed by "Fresh-water-Fish", a short list of 4 species with some notes occupying half a page.

Heading: The Names of Plants in Dale's Pharmacologia, with Reference to their pages.
Pages: Sixty-five, many blank.
Notes: This is a partial index to Samuel Dale, *Pharmacologia* (1693). It relates only to the "Phytologia seu Materiae Medicae pars Secunda" (pages 102–523 of the *Pharmacologia*). Other sections of the work deal with stones, minerals and chemicals (part 1), and with animal compounds (part 3). Nicolson started to make an index to the botanical part and abandoned it at page 181, possibly after the 1705 supplement to the *Pharmacologia* came out with an English names glossary.

Sixteen blank pages then follow, and the next items read from the end of the volume, which has been turned upside-down. On the final page is written "H. Cotton." and there is a jotting of a plant with location: "Alsine Tenuifolia. J.B. by the Maiden pink at Common-holm bridge." (This is the polynomial for *Minuartia hybrida.* It is probably a mis-identification for *M. verna*).

There follows the garden list printed in full below.

Sett in my own Garden Sep. 1690. By Mr Lawson etc.

Under the West-Wall –
1. Staphylodendron.
2. Pinus.
3. Cypressus.
4. Juniperus.
5. Colutea Scorpioides.

On the border opposite –
1. Syringa Alba.
2. —— Coerulea.
3. Horse Chesnut.
4. Black Cherry.
5. Pseudo-Cytisus.
6. Pentaphylloides fruticosum.

In two or three adjoyning Beds –
1. Filipendula.
2. Rhodia Radix.
3. Polygonatum.
4. Asarabacca.
5. Sanicula Alpina guttata.
6. Verbena.
7. Calamintha mont*ana* praest*antior* Elegans.
8. Doronicon plantaginis folio.
9. Geranium nodosum.
10. Ger*anium* Haematodes flore eleg*anter* variegato.
11. Aster Virgineanus Luteus.
12. Pilosella maxima.
13. Campanula Urticae folio.
14. Martagon.
15. Cardiaca.
16. Mentastrum folio rug*oso* rotundo.
17. Valeriana Graeca.
18. Valeriana Dodonaei.
19. Lychnis Constantinop*olitana*.
20. Thalictrum majus.
21. Hedera Virginiana trifolia.
22. Alchymilla Alpina pentaph*yllos*.
23. Lysimachia lutea.
24. Androsaemum.
25. Parietaria.
26. Melissa.
27. Cyanus Babylonicus.

28. Chamaedrys Hortensis.
29. Scordium.
30. Telephium.
31. Althaea.
32. Dentariae affinis.

In the 2d. Bed. –

1. J*u*nc*u*s Graminea.
2. Gramen Striatum.
3. Imperatoria.
4. Borrago semper vivens.
5. Symphytum.
6. Scelery.
7. Leucoium.
8. Rubarb.
9. Marrubium Aquaticum.
10. Ageratum Hortense.
11. Pionia mas.
12. Geranium moschatum.
13. Mentha Crispa.
14. Muscipula Lobelij.
15. Terragon.
16. Alcacangi.

In the 3d Bed –

1. Viola Lunaris.
2. Herba Gerardi.
3. Herba Doria.
4. Chamaenerion. Lysim*achia* Speciosa.
5. Lepidium.

On the Border above the green plott –

1. Euonymus Theophrasti.
2. Filberts.
3. Populus Alba.
4. Cornus foemina.
5. Syringa Coerulea.
6. Spiraea Theophrasti.
7. Solanum Lignosum.
8. Tilia.

On the little border at the end –

1. Pentaphyl*loides* fruticosum.
2. Spiraea Theophrasti.

On the great border below the plott –

1. Rosa lutea simplex.
2. Draba Lutea siliq*uis* strictiss*imis*.
3. Doronicum Laciniatum.
4. Campanula Urticae folio.
5. Cornus foemina.
6. Pseudo-Cytisus.

Under the House-Wall –

1. Jesmina.
2. Ficus.

On a small ⟨border⟩[98] by the Door –

1. Hepatica.
2. Calceolus Mariae.
3. Lilium Album.
4. Colchicum.
5. Armerius repens.
6. Rhodia radix.
7. Lilium Convallium.

At the end of this list is a note: "Form'd Stones (22 sorts) sent from Mr Lhwyd. May, 30. 93. See my Almanack for that year."

[Lists and Notes]

This is followed by a page of jottings of receipts by Nicolson (July 1695–June 1696) of some fossils and antiquities.

Heading: Musaeum Physicum.
Pages: Four.
Notes: A list of geological specimens.

There follows two pages of receipts by Nicolson (June 1696–March 1697) of old coins.

Lastly there are two more garden lists dating from his instalment at Rose in 1703. They are printed in full below.

98. Supplied by analogy with previous usage.

Nov. 19. 1703

Set in the Garden at Rose (and my wife's little Garden) the following Shrubs, sent from Mr Sutherland; Or, at least, so many of 'em as the Carrier brought to my hand.

1. Deep purple Lilac.
2. Dwarf Medlar.
3. White Beam Tree.
4. Rose without Thorns.
5. Yellew (sic) Rose.
6. Buck ⟨t⟩ horn-Tree.
7. French Tamarisk.
8. German Tamarisk.
9. Sea Buck-Thorn.
10. White flower'd Lilac.
11. Mock Willow.
12. Upright Honey-suckle.
13. Wayfareing Tree.
14. Shrub Trefoil.
15. Early-flowring upright Honysuckle of the Alps.
16. Late-flowring upright Honysuckle of the Alps.
17. Dwarf Almond.
18. Sweet-smelling American Rasp.
19. Shrub St Johnswort.
20. Curran with Goosberry Leaves.
21. Persian Jasmin.
22. White Pipe Tree.

Mar. 24. 1703/4. Set at Rose, from Mr Sutherland –

1. Ranunculus Lanuginosus grumosa radice. *Illyrian Crowfoot.*
2. Chondrilla Viminea. *Twig-branch'd Gum-Succory.*
3. Clematis Pannonica. *Hungarian Climmer.*
4. Veronica erecta quadrifolia Virginiana. *Virginian Speedwell.*
5. Eryngium Mediterraneum. *Land-Eryngo.*
6. Aconitum flore ex albo et coeruleo variegato. *Party-colour'd Monk Hood.*
7. Chamaedrys Teucrij facie. *Mock Tree-Germander.*
8. Cynoglossum Creticum. *Hounds-Tongue of Candie.*
9. Papaver Pyraenaicum flore luteo. *Mountain-poppy with a yellow flower.*
10. Pulmonaria maculosa flore albo. *Spotted Lungwort with a white flower.*
11. Iris Gloriosa. *Proud Flower de Lis.*
12. Jacobaea Maritima folijs minus incanis. *Sea Ragwort.*
13. Aster Virginianus Serotinus flore parvo albente. *Narrow-leav'd Virginian Starwort.*
14. Telephium minus repens folijs deciduis. *Creeping Orpine.*
15. Ptarmica Matricariae folijs. *Feverfew-leav'd Sneezewort.*
16. Androsaemum flore maximo Montis OLympi. *Largeflower'd Tutsan.*
17. Iris folio Gramineo. [Grass]
18. Hepatica Nobilis flore rubro.
19. Ranunculus montanus folio gramineo. *Mountain-Crowfoot.*
20. Cortusa Matthioli. *Bears-ear Sanicle.*
21. Ranunculus folio plantaginis.
22. Absinthium maritimum Lavendulae folio. *Sea Mugwort.*
23. Chrysanthemum erectum Angustifolium Virginianum altissimum. *Virginian Corn-Marygold.*
24. Millefolium maximum flore albo.
25. Valeriana Alliariae folio.
26. Linaria purpurea odorata.
27. Aster Novae Angliae Latifolius paniculatus flore saturate violaceo.
28. Thlaspi Creticum semper Virens. *Ever-green Candy-Tufts.*
29. Aster montanus serotinus folijs subrotundis. *Late Mountain Starwort.*
30. Thalictrum Canadense Aquilegiae folijs. *American Meadow-Rue.*
31. Doronicum Americanum. *American Leopards-Bane.*
32. Aster Tripolij flore Latifolius. *Broad leav'd Starwort.*
33. Aconitum Serotinum flore coeruleo. *Lateflowring Monkshood.*
34. Cortusa Americana flore squallide purpureo. *American Sanicle.* Qu.
35. Trifolium Acetosum erectum luteum Virginianum. *Virginia-Woodsorrel.*

36. Origanum folijs ex albo variegatis. *party-colour'd Wild Marjoram.*
37. Origanum folijs ex luteo variegatis.
38. Flos solis radice perenni. *Perennial Sun-flower.*
39. Scabiosa Argentea Angustifolia. *Silver-leav'd Scabious.*
40. Geranium tuberosa radice. *Bulbous rooted Cranesbill.*

. . .

GAZETTEER

OUTSIDE CUMBRIA

REFERENCE

-·-·-·- County boundaries of Cumberland, Westmorland and Furness as in Nicolson's day, which, with the former rural district of Sedbergh, now form the County of Cumbria.

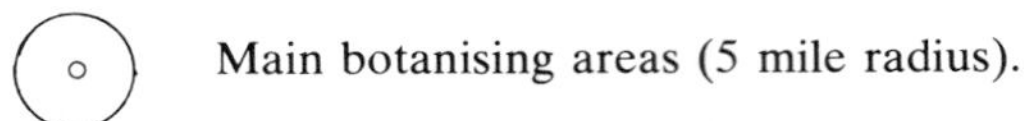

Main botanising areas (5 mile radius).

• Plant locations outside main areas. Where more than one plant is recorded, the number is shown.

National grid references are given in the text. Marginal ticks at 10 kilometre intervals are shown on the map.

NUMBERS OF RECORDS

	Flowering	*Non-flowering*
In Main Botanising area		
Carlisle	22	
Parish of Great Salkeld	350	30
Great Strickland	60	2
Shap Fells	16	7
Kendal	62	5
Walney Island	25	2
Elsewhere in Cumbria	117	9
Total records in Cumbria	652	55

Some plants have been recorded in more than one area: Adjusted total Cumbria

553 Flowering
48 Non-flowering

601

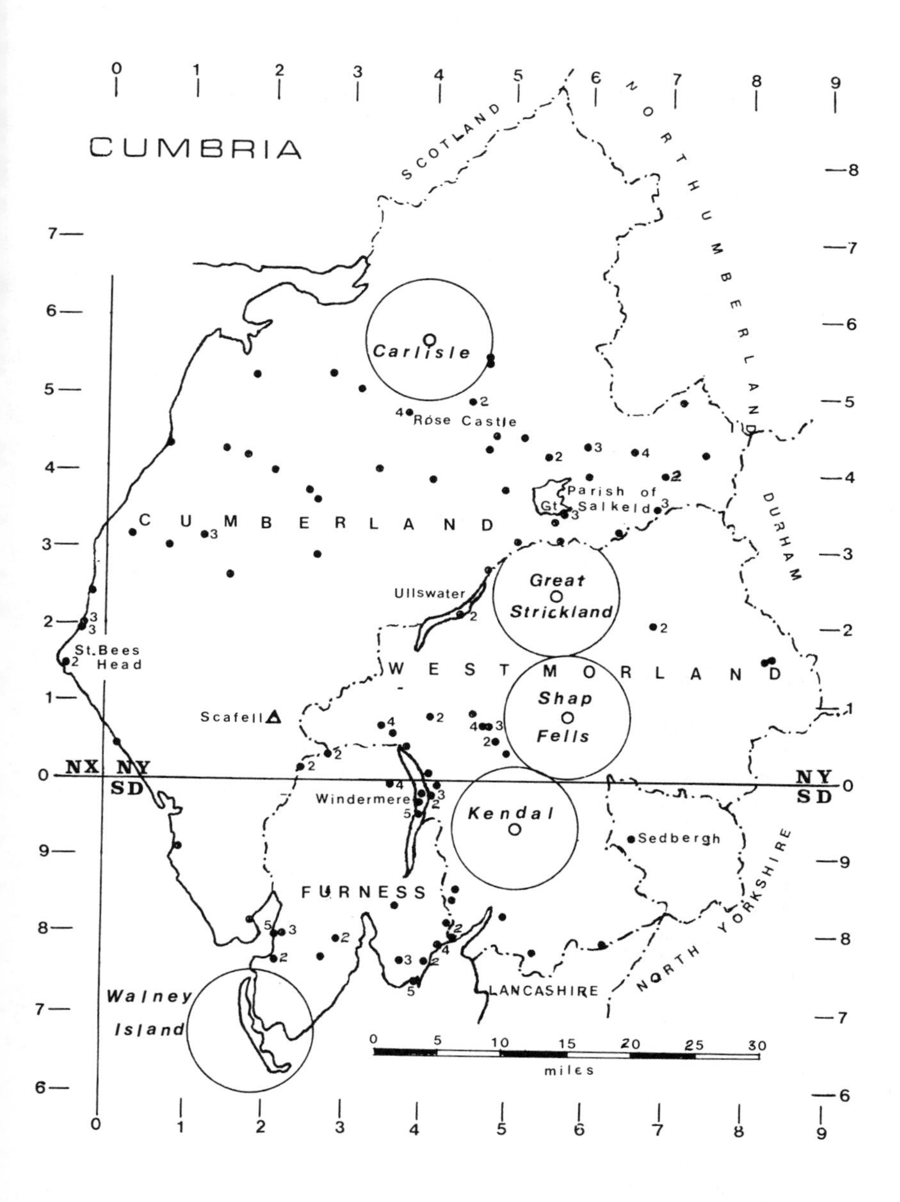
CUMBRIA
SCOTLAND
NORTHUMBERLAND
DURHAM
CUMBERLAND
WESTMORLAND
FURNESS
LANCASHIRE
NORTH YORKSHIRE
Carlisle
Rose Castle
Parish of Gt. Salkeld
Great Strickland
Ullswater
Shap Fells
Kendal
Windermere
Sedbergh
St. Bees Head
Scafell
Walney Island
NX NY SD
NY SD
miles

INDEX OF MODERN BOTANICAL NAMES